I0887360

AUTONOMY RESEARCH FOR CIVIL AVIATION

TOWARD A NEW ERA OF FLIGHT

Committee on Autonomy Research for Civil Aviation

Aeronautics and Space Engineering Board

Division on Engineering and Physical Sciences

NATIONAL RESEARCH COUNCIL
OF THE NATIONAL ACADEMIES

THE NATIONAL ACADEMIES PRESS
Washington, D.C.
www.nap.edu

THE NATIONAL ACADEMIES PRESS **500 Fifth Street, NW** **Washington, DC 20001**

NOTICE: The project that is the subject of this report was approved by the Governing Board of the National Research Council, whose members are drawn from the councils of the National Academy of Sciences, the National Academy of Engineering, and the Institute of Medicine. The members of the committee responsible for the report were chosen for their special competences and with regard for appropriate balance.

This report is based on work supported by Contract NNH10CD04B between the National Academy of Sciences and the National Aeronautics and Space Administration. Any opinions, findings, conclusions, or recommendations expressed in this publication are those of the authors and do not necessarily reflect the views of the agency that provided support for the project.

International Standard Book Number-13: 978-0-309-30614-0
International Standard Book Number-10: 0-309-30614-0

Cover: Design by Tim Warchocki.

Copies of this report are available free of charge from the

Aeronautics and Space Engineering Board
National Research Council
Keck Center of the National Academies
500 Fifth Street, NW
Washington, DC 20001

Additional copies of this report are available from the

National Academies Press
Keck 360
500 Fifth Street, NW
Washington, DC 20001
(800) 624-6242 or (202) 334-3313
http://www.nap.edu.

Copyright 2014 by the National Academy of Sciences. All rights reserved.

Printed in the United States of America

THE NATIONAL ACADEMIES

Advisers to the Nation on Science, Engineering, and Medicine

The **National Academy of Sciences** is a private, nonprofit, self-perpetuating society of distinguished scholars engaged in scientific and engineering research, dedicated to the furtherance of science and technology and to their use for the general welfare. Upon the authority of the charter granted to it by the Congress in 1863, the Academy has a mandate that requires it to advise the federal government on scientific and technical matters. Dr. Ralph J. Cicerone is president of the National Academy of Sciences.

The **National Academy of Engineering** was established in 1964, under the charter of the National Academy of Sciences, as a parallel organization of outstanding engineers. It is autonomous in its administration and in the selection of its members, sharing with the National Academy of Sciences the responsibility for advising the federal government. The National Academy of Engineering also sponsors engineering programs aimed at meeting national needs, encourages education and research, and recognizes the superior achievements of engineers. Dr. C. D. Mote, Jr., is president of the National Academy of Engineering.

The **Institute of Medicine** was established in 1970 by the National Academy of Sciences to secure the services of eminent members of appropriate professions in the examination of policy matters pertaining to the health of the public. The Institute acts under the responsibility given to the National Academy of Sciences by its congressional charter to be an adviser to the federal government and, upon its own initiative, to identify issues of medical care, research, and education. Dr. Victor J. Dzau is president of the Institute of Medicine.

The **National Research Council** was organized by the National Academy of Sciences in 1916 to associate the broad community of science and technology with the Academy's purposes of furthering knowledge and advising the federal government. Functioning in accordance with general policies determined by the Academy, the Council has become the principal operating agency of both the National Academy of Sciences and the National Academy of Engineering in providing services to the government, the public, and the scientific and engineering communities. The Council is administered jointly by both Academies and the Institute of Medicine. Dr. Ralph J. Cicerone and Dr. C. D. Mote, Jr., are chair and vice chair, respectively, of the National Research Council.

www.nationalacademies.org

OTHER REPORTS OF THE AERONAUTICS AND SPACE ENGINEERING BOARD

Continuing Kepler's Quest: Assessing Air Force Space Command's Astrodynamics Standards (Aeronautics and Space Engineering Board [ASEB], 2012)
NASA Space Technology Roadmaps and Priorities: Restoring NASA's Technological Edge and Paving the Way for a New Era in Space (ASEB, 2012)
NASA's Strategic Direction and the Need for a National Consensus (Division on Engineering and Physical Sciences, 2012)
Recapturing NASA's Aeronautics Flight Research Capabilities (Space Studies Board [SSB] and ASEB, 2012)
Reusable Booster System: Review and Assessment (ASEB, 2012)
Solar and Space Physics: A Science for a Technological Society (SSB with ASEB, 2012)

Limiting Future Collision Risk to Spacecraft: An Assessment of NASA's Meteroid and Orbital Debris Programs (ASEB, 2011)
Preparing for the High Frontier—The Role and Training of NASA Astronauts in the Post-Space Shuttle Era (ASEB, 2011)

Advancing Aeronautical Safety: A Review of NASA's Aviation Safety-Related Research Programs (ASEB, 2010)
Capabilities for the Future: An Assessment of NASA Laboratories for Basic Research (Laboratory Assessments Board with SSB and ASEB, 2010)
Defending Planet Earth: Near-Earth Object Surveys and Hazard Mitigation Strategies (SSB with ASEB, 2010)
Forging the Future of Space Science: The Next 50 Years: An International Public Seminar Series Organized by the Space Studies Board: Selected Lectures (SSB with ASEB, 2010)
Life and Physical Sciences Research for a New Era of Space Exploration: An Interim Report (SSB with ASEB, 2010)
Recapturing a Future for Space Exploration: Life and Physical Sciences Research for a New Era (ASEB, 2010)

America's Future in Space: Aligning the Civil Space Program with National Needs (SSB with ASEB, 2009)
Approaches to Future Space Cooperation and Competition in a Globalizing World: Summary of a Workshop (SSB with ASEB, 2009)
An Assessment of NASA's National Aviation Operations Monitoring Service (ASEB, 2009)
Final Report of the Committee for the Review of Proposals to the 2009 Engineering and Physical Science Research and Commercialization Program of the Ohio Third Frontier Program (ASEB, 2009)
Fostering Visions for the Future: A Review of the NASA Institute for Advanced Concepts (ASEB, 2009)
Near-Earth Object Surveys and Hazard Mitigation Strategies: Interim Report (SSB with ASEB, 2009)
Radioisotope Power Systems: An Imperative for Maintaining U.S. Leadership in Space Exploration (SSB with ASEB, 2009)

Limited copies of ASEB reports are available free of charge from

Aeronautics and Space Engineering Board
National Research Council
Keck Center of the National Academies
500 Fifth Street, NW, Washington, DC 20001
(202) 334-2858/aseb@nas.edu
www.nationalacademies.org/aseb

COMMITTEE ON AUTONOMY RESEARCH FOR CIVIL AVIATION

JOHN-PAUL B CLARKE, Georgia Institute of Technology, *Co-chair*
JOHN K. LAUBER, Consultant, *Co-chair*
BRENT APPLEBY, Draper Laboratory
ELLA M. ATKINS, University of Michigan
ANTHONY J. BRODERICK, Consultant
GARY L. COWGER, GLC Ventures, LLC
CHRISTOPHER E. FLOOD, Delta Air Lines
MICHAEL S. FRANCIS, United Technologies Research Center
ERIC FREW, University of Colorado, Boulder
ANDREW LACHER, MITRE Corporation
JOHN D. LEE, University of Wisconsin-Madison
KENNETH M. ROSEN, General Aero-Science Consultants, LLC
LAEL RUDD, Northrop Grumman Aerospace Systems
PATRICIA VERVERS, Honeywell Aerospace
LARRELL B. WALTERS, University of Dayton Research Institute
DAVID D. WOODS, Ohio State University
EDWARD L. WRIGHT, University of California, Los Angeles

Staff

ALAN C. ANGLEMAN, Senior Program Officer, *Study Director*
MICHAEL H. MOLONEY, Director, Aeronautics and Space Engineering Board and Space Studies Board
LEWIS GROSWALD, Associate Program Officer
LINDA WALKER, Senior Program Assistant
ANESIA WILKS, Program Assistant

AERONAUTICS AND SPACE ENGINEERING BOARD

LESTER L. LYLES, The Lyles Group, *Chair*
PATRICIA GRACE SMITH, Aerospace Consultant, Washington, D.C., *Vice Chair*
ARNOLD D. ALDRICH, Aerospace Consultant, Vienna, Virginia
ELLA M. ATKINS, University of Michigan
STEVEN J. BATTEL, Battel Engineering
BRIAN J. CANTWELL, Stanford University
ELIZABETH R. CANTWELL, Lawrence Livermore National Laboratory
EILEEN M. COLLINS, Space Presentations, LLC
RAVI B. DEO, President, EMBR
VIJAY K. DHIR, University of California at Los Angeles
EARL H. DOWELL, Duke University
ALAN H. EPSTEIN, Technology & Environment, Pratt & Whitney
KAREN FEIGH, Georgia Tech College of Engineering
PERETZ P. FRIEDMANN, University of Michigan
MARK J. LEWIS, IDA Science and Technology Policy Institute
JOHN M. OLSON, Space Systems Group, Sierra Nevada Corporation
HELEN L. REED, Texas A & M University
AGAM N. SINHA, ANS Aviation International, LLC
JOHN P. STENBIT, Consultant, Oakton, Virginia
ALAN M. TITLE, Advanced Technology Center, Lockheed Martin
DAVID M. VAN WIE, Applied Physics Laboratory, Johns Hopkins University

Staff

MICHAEL H. MOLONEY, Director
CARMELA J. CHAMBERLAIN, Administrative Coordinator
TANJA PILZAK, Manager, Program Operations
CELESTE A. NAYLOR, Information Management Associate
CHRISTINA O. SHIPMAN, Financial Officer
SANDRA WILSON, Financial Assistant

Preface

Technological advances in computer systems, sensors, precision position and navigation information, and other areas are facilitating the development and operation of increasingly autonomous (IA) systems and vehicles for a wide variety of applications on the ground, in space, at sea, and in the air. IA systems have the potential to improve safety and reliability, reduce costs, and enable new missions. However, deploying IA systems is not without risk. In particular, failure to implement IA systems in a careful and deliberate manner could potentially reduce safety and/or reliability and increase life-cycle costs. These factors are especially critical to civil aviation given the very high standards for safety and reliability and the risk to public safety that occurs whenever the performance of new civil aviation technologies or systems falls short of expectations.

Research and technology development plays a critical role in determining the effectiveness of IA systems and the pace at which they advance. In addition, a wide variety of organizations possess key expertise and are making advances in technologies directly related to the advancement of IA systems for civil aviation. Accordingly, NASA's Aeronautics Research Mission Directorate requested that the NRC convene a committee to develop a *national* research agenda for autonomy in civil aviation. In response, the Aeronautics and Space Engineering Board of the Division on Engineering and Physical Sciences, with the assistance of the Board on Human–Systems Integration of the Division of Behavioral and Social Sciences and Education, assembled a committee to carry out the assigned statement of task. As specified in that statement, the committee developed a research agenda consisting of a prioritized set of research projects that, if completed by NASA and other interested parties, would enable concepts of operation for the National Airspace System whereby ground systems and aircraft with various autonomous capabilities would be able to operate in harmony; demonstrate IA capabilities for crewed and unmanned aircraft; predict the system-level effects of incorporating IA systems and aircraft in the National Airspace System; and define approaches for verification, validation, and certification of IA systems. The committee was also tasked with describing contributions that advances in autonomy could make to civil aviation and the technical and policy barriers that must be overcome to fully and effectively implement IA systems in civil aviation.

Unlike typical NRC reports, this report often cites news media (magazines, newspapers, and online blogs) as the source of information contained in the report. In all cases, these sources are used to report recent events in the fast-changing world of IA systems and unmanned aircraft. They are not used as the basis for any scientific or technical analysis or conclusions.

The staff of the Board on Human-Systems Integration assisted the staff of the Aeronautics and Space Engineering Board in developing the statement of task for this study and in forming the 17-member committee. The committee met five times during 2013 and 2014: three times in Washington, D.C., and twice in Irvine, California. At these meetings the committee was informed by presentations and materials provided by current and former personnel from the Federal Aviation Administration, NASA, the National Highway Traffic Safety Administration, the U.S. Air Force, the U.S. Army, the U.S. Navy, academia, industry, and independent research institutes.

John-Paul Clarke, *Co-chair*
John Lauber, *Co-chair*
Committee on Autonomy Research for Civil Aviation

Acknowledgment of Reviewers

This report has been reviewed in draft form by individuals chosen for their diverse perspectives and technical expertise, in accordance with procedures approved by the National Research Council's (NRC's) Report Review Committee. The purpose of this independent review is to provide candid and critical comments that will assist the institution in making its published report as sound as possible and to ensure that the report meets institutional standards for objectivity, evidence, and responsiveness to the study charge. The review comments and draft manuscript remain confidential to protect the integrity of the deliberative process. We wish to thank the following individuals for their review of this report:

Timothy J. Buker, Gulfstream Aerospace Corporation,
Renwick E. Curry, University of California, Santa Cruz,
George L. Donohue, George Mason University (emeritus),
Neil Gehrels, NASA Goddard Space Flight Center,
John R. Huff, Oceaneering International, Inc.,
Charles Lee Isbell, Jr., Georgia Institute of Technology,
Vijay Kumar, University of Pennsylvania,
David Mindell, Massachusetts Institute of Technology,
Agam N. Sinha, ANS Aviation International, LLC,
George W. Swenson, Jr., University of Illinois, Urbana-Champaign, and
James M. Wasiloff, U.S. Army, Tank-Automotive and Armaments Command.

Although the reviewers listed above have provided many constructive comments and suggestions, they were not asked to endorse the conclusions or recommendations, nor did they see the final draft of the report before its release. The review of this report was overseen by Marcia Rieke, University of Arizona, and John Klineberg, Space Systems/Loral (retired). Appointed by the NRC, they were responsible for making certain that an independent examination of this report was carried out in accordance with institutional procedures and that all review comments were carefully considered. Responsibility for the final content of this report rests entirely with the authoring committee and the institution.

Contents

Summary

The development and application of increasingly autonomous (IA) systems for civil aviation (see Boxes S.1 and S.2) are proceeding at an accelerating pace, driven by the expectation that such systems will return significant benefits in terms of safety, reliability, efficiency, affordability, and/or previously unattainable mission capabilities. IA systems, characterized by their ability to perform more complex mission-related tasks with substantially less human intervention for more extended periods of time, sometimes at remote distances, are being envisioned for aircraft and for air traffic management (ATM) and other ground-based elements of the National Airspace System (NAS) (see Box S.3). This vision and the associated technological developments have been spurred in large part by the convergence of the increased availability of low-cost, highly capable computing systems; sensor technologies; digital communications systems; precise position, navigation, and timing information (e.g., from the Global Positioning System (GPS); and open-source hardware and software.

These technology enablers, coupled with expanded use of IA systems in military operations and the emergence of an active and growing community of hobbyists that is developing and operating small unmanned aircraft systems (UAS), provide fertile ground for innovation and entrepreneurship (see Box S.4). The burgeoning industrial sector devoted to the design, manufacture, and sales of IA systems is indicative of the perceived economic opportunities that will arise. In short, civil aviation is on the threshold of potentially revolutionary changes in aviation capabilities and operations associated with IA systems. These systems, however, pose serious unanswered questions about how to safely integrate these revolutionary technological advances into a well-established, safe, and efficiently functioning NAS governed by operating rules that can only be changed after extensive deliberation and consensus. In addition, the potential benefits that could accrue from the introduction of advanced IA systems in civil aviation, the associated costs, and the unintended consequences that are likely to arise will not fall on all stakeholders equally. This report suggests major elements of a national research agenda for autonomy in civil aviation that would inform and support the orderly implementation of IA systems in U.S. civil aviation. The scope of this study does not include organizational recommendations.

BARRIERS TO INCREASED AUTONOMY IN CIVIL AVIATION

The Committee on Autonomy Research for Civil Aviation has identified many substantial barriers to the increased use of autonomy in civil aviation systems and aircraft. These barriers cover a wide range of issues related

BOX S.1
Civil Aviation

In this report, "civil aviation" is used to refer to all nonmilitary aircraft operations in U.S. civil airspace. This includes operations of civil aircraft as well as nonmilitary public use aircraft (that is, aircraft owned or operated by federal, state, and local government agencies other than the Department of Defense). In addition, many of the IA technologies that would be developed by the recommended research projects would generally be applicable to military crewed and/or unmanned aircraft for military operations and/or other operations in the NAS.

BOX S.2
Increasingly Autonomous Systems

A fully autonomous aircraft would not require a pilot; it would be able to operate independently within civil airspace, interacting with air traffic controllers and other pilots just as if a human pilot were on board and in command. Similarly, a fully autonomous ATM system would not require human air traffic controllers. This study is not focused on these extremes (although it does sometimes address the needs or qualities of fully autonomous unmanned aircraft). Rather, the report primarily addresses what the committee calls "increasingly autonomous" (IA) systems, which lie along the spectrum of system capabilities that begin with the abilities of current automatic systems, such as autopiloted and remotely piloted (nonautonomous) unmanned aircraft, and progress toward the highly sophisticated systems that would be needed to enable the extreme cases. Some IA systems, particularly adaptive/nondeterministic IA systems, lie farther along this spectrum than others, and in this report such systems are typically described as "advanced IA systems."

BOX S.3
National Airspace System

The NAS is "the common network of U.S. airspace; air navigation facilities, equipment, and services; airports or landing areas; aeronautical charts, information and services; rules, regulations, and procedures; technical information; and manpower and material" (Integration of Civil Unmanned Aircraft Systems [UAS] in the National Airspace System [NAS] Roadmap, FAA, 2013). Some NAS facilities are jointly operated by the FAA and the Department of Defense. IA systems could be incorporated into airport ground systems such as snow plows. However, the greatest technological, social, and legal challenges to the use of IA systems in civil aviation are associated with their use in aircraft and air traffic management systems, and the report does not specifically address the use of IA systems in airport ground systems.

BOX S.4
Unmanned Aircraft/Crewed Aircraft

An unmanned aircraft is "a device used or intended to be used for flight in the air that has no onboard pilot. This device excludes missiles, weapons, or exploding warheads, but includes all classes of airplanes, helicopters, airships, and powered-lift aircraft without an onboard pilot. Unmanned aircraft do not include traditional balloons (see 14 CFR Part 101), rockets, tethered aircraft and un-powered gliders." A UAS is "an unmanned aircraft and its associated elements related to safe operations, which may include control stations (ground-, ship-, or air-based), control links, support equipment, payloads, flight termination systems, and launch/recovery equipment" (Integration of Civil Unmanned Aircraft Systems [UAS] in the National Airspace System [NAS] Roadmap, Federal Aviation Administration [FAA], 2013). UAS include the data links and other communications systems used to connect the UAS control station, unmanned aircraft, and other elements of the NAS, such as ATM systems and human operators. Unless otherwise specified, UAS are assumed to have no humans on board either as flight crew or as passengers. "Crewed aircraft" is used to denote manned aircraft; unless specifically noted otherwise; manned aircraft are considered to have a pilot on board.

to understanding, developing, and deploying IA ground and aircraft systems. Some of these issues are technical, some are related to certification and regulation, and some are related to legal and social concerns.

- Technology barriers
  —*Communications and data acquisition.* Civil aviation wireless communications are fundamentally limited in bandwidth, and the operation of unmanned aircraft in the NAS could substantially increase the demand for bandwidth.
  —*Cyberphysical security.* The use of increasingly interconnected networks and increasingly complex software embedded throughout IA air- and ground-based system elements, as well as the increasing sophistication of potential cyberphysical attacks, threaten the safety and reliability of IA systems.
  —*Decision making by adaptive/nondeterministic systems* (see Box S.5). The lack of generally accepted design, implementation, and test practices for adaptive/nondeterministic systems will impede the deployment of some advanced IA vehicles and systems in the NAS.
  —*Diversity of vehicles.* It will be difficult to engineer some IA systems so that they are backward-compatible with legacy airframes, ATM systems, and other elements of the NAS.
  —*Human–machine integration.* Incorporating IA systems and vehicles in the NAS would require humans and machines to work together in new and different ways that have not yet been identified.
  —*Sensing, perception, and cognition.* The ability of IA systems to operate independently of human operators (see Box S.6) is fundamentally limited by the capabilities of machine sensory, perceptual, and cognitive systems.
  —*System complexity and resilience.* IA capabilities create a more complex aviation system, with new interdependencies and new relationships among various operational elements. This will likely reduce the resilience of the NAS because disturbances in one portion of the system could, in certain circumstances, cause the performance of the entire system to degrade precipitously.
  —*Verification and validation (V&V).* Existing V&V approaches and methods are insufficient for advanced IA systems.
- Regulation and certification barriers
  —*Airspace access for unmanned aircraft.* Unmanned aircraft may not operate in nonsegregated civil airspace unless the Federal Aviation Administration (FAA) issues a certificate of waiver or authorization (COA).

BOX S.5
Adaptive/Nondeterministic Systems

Adaptive systems have the ability to modify their behavior in response to their external environment. For aircraft systems, this could include commands from the pilot and inputs from aircraft systems, including sensors that report conditions outside the aircraft. Some of these inputs, such as airspeed, will be stochastic because of sensor noise as well as the complex relationship between atmospheric conditions and sensor readings not fully captured in calibration equations. Adaptive systems learn from their experience, either operational or simulated, so that the response of the system to a given set of inputs varies and, presumably, improves over time.

Systems that are nondeterministic may or may not be adaptive. They may be subject to the stochastic influences imposed by their complex internal operational architectures or their external environment, meaning that they will not always respond in precisely the same way even when presented with identical inputs or stimuli. The software that is at the heart of nondeterministic systems is expected to enable improved performance because of its ability to manage and interact with complex "world models" (large and potentially distributed data sets) and execute sophisticated algorithms to perceive, decide, and act in real time.

Systems that are adaptive and nondeterministic demonstrate the performance enhancements of both. Many advanced IA systems are expected to be adaptive and/or nondeterministic, and issues associated with the development and deployment of these adaptive/nondeterministic systems are discussed later in the report.

BOX S.6
Operators

In this report, the term "operator" generally refers to pilots, air traffic controllers, airline flight operations staff, and other personnel who interact directly with IA civil aviation systems. "Pilot" is used when referring specifically to the operator of a crewed aircraft. With regard to unmanned aircraft, the FAA says that "in addition to the crewmembers identified in 14 CFR Part 1 [pilots, flight engineers, and flight navigators], a UAS flight crew includes pilots, sensor/payload operators, and visual observers, but may include other persons as appropriate or required to ensure safe operation of the aircraft" (Integration of Civil Unmanned Aircraft Systems [UAS] in the National Airspace System [NAS] Roadmap, FAA, 2013). Given that the makeup, certification requirements, and roles of UAS flight crews are likely to evolve as UAS acquire advanced IA capabilities, this report refers generally to UAS operators as flight crew rather than specifically as pilots.

—*Certification process.* Existing certification criteria, processes, and approaches do not take into account the special characteristics of advanced IA systems.

—*Equivalent level of safety.* Many existing safety standards and requirements, which are focused on assuring the safety of aircraft passengers and crew on a particular aircraft, are not well suited to assure the safety of unmanned aircraft operations, where the primary concern is the safety of personnel in other aircraft and on the ground.

—*Trust in adaptive/nondeterministic IA systems.* Verification, validation, and certification are necessary but not sufficient to engender stakeholder trust in advanced adaptive/nondeterministic IA systems.

- Other barriers
  —*Legal issues.* Public policy, as reflected in law and regulation, could significantly impede the degree and speed of adoption of IA technology in the NAS.
  —*Social issues.* Social issues, particularly public concerns about privacy and safety, could significantly impede the degree and speed of adoption of IA technology in the NAS.

The committee did not individually prioritize these barriers. However, there is one critical, crosscutting challenge that must be overcome to unleash the full potential of advanced IA systems in civil aviation. This challenge may be described in terms of a question: "How can we assure that advanced IA systems—especially those systems that rely on adaptive/nondeterministic software—will enhance rather than diminish the safety and reliability of the NAS?" There are four particularly challenging barriers that stand in the way of meeting this critical challenge:

- Certification process
- Decision making by adaptive/nondeterministic systems
- Trust in adaptive/nondeterministic IA systems
- Verification and validation

ELEMENTS OF A NATIONAL RESEARCH AGENDA FOR AUTONOMY IN CIVIL AVIATION

The committee identified eight high-level research projects that would address the barriers discussed above. The committee also identified several specific areas of research that could be included in each research project.

Recommendation. *National Research Agenda.* Agencies and organizations in government, industry, and academia that are involved in research, development, manufacture, certification, and regulation of IA technologies and systems should execute a national research agenda in autonomy that includes the following high-priority research projects, with the first four being the most urgent and the most difficult:

- ***Behavior of Adaptive/Nondeterministic Systems.* Develop methodologies to characterize and bound the behavior of adaptive/nondeterministic systems over their complete life cycle.**
- ***Operation Without Continuous Human Oversight.* Develop the system architectures and technologies that would enable increasingly sophisticated IA systems and unmanned aircraft to operate for extended periods of time without real-time human cognizance and control.**
- ***Modeling and Simulation.* Develop the theoretical basis and methodologies for using modeling and simulation to accelerate the development and maturation of advanced IA systems and aircraft.**
- ***Verification, Validation, and Certification.* Develop standards and processes for the verification, validation, and certification of IA systems, and determine their implications for design.**
- ***Nontraditional Methodologies and Technologies.* Develop methodologies for accepting technologies not traditionally used in civil aviation (e.g., open-source software and consumer electronic products) in IA systems.**
- ***Roles of Personnel and Systems.* Determine how the roles of key personnel and systems, as well as related human–machine interfaces, should evolve to enable the operation of advanced IA systems.**
- ***Safety and Efficiency.* Determine how IA systems could enhance the safety and efficiency of civil aviation.**
- ***Stakeholder Trust.* Develop processes to engender broad stakeholder trust in IA systems for civil aviation.**

FOUR MOST URGENT AND MOST DIFFICULT RESEARCH PROJECTS

Behavior of Adaptive/Nondeterministic Systems. Develop methodologies to characterize and bound the behavior of adaptive/nondeterministic systems over their complete life cycle.

Adaptive/nondeterministic properties will be integral to many advanced IA systems, but they will create challenges for assessing and setting the limits of their resulting behaviors. Advanced IA systems for civil aviation operate in an uncertain environment where physical disturbances, such as wind gusts, are often modeled using probabilistic models. These IA systems may rely on distributed sensor systems that have noise with stochastic properties such as uncertain biases and random drifts over time and varying environmental conditions. To improve performance, adaptive/nondeterministic IA systems will take advantage of evolving conditions and past experience to adapt their behavior; that is, they will be capable of learning. As these IA systems take over more functions traditionally performed by humans, there will be a growing need to incorporate autonomous monitoring and other safeguards to ensure continued appropriate operational behavior.

There is tension between the benefits of incorporating software with adaptive/nondeterministic properties in IA systems and the requirement to test such software for safe and assured operation. Research is needed to develop new methods and tools to address the inherent uncertainties in airspace system operations and thereby enable more complex adaptive/nondeterministic IA systems with the ability to adapt over time to improve their performance and provide greater assurance of safety.

Specific tasks to be carried out by this research project include the following:

- Develop mathematical models for describing adaptive/nondeterministic processes as applied to humans and machines.
- Develop performance criteria, such as stability, robustness, and resilience, for the analysis and synthesis of adaptive/nondeterministic behaviors.
- Develop methodologies beyond input-output testing for characterizing the behavior of IA systems.
- Determine the roles that humans play in limiting the behavior of adaptive/nondeterministic systems and how IA systems can take over those roles.

Operation Without Continuous Human Oversight. Develop the system architectures and technologies that would enable increasingly sophisticated IA systems and unmanned aircraft to operate for extended periods of time without real-time human cognizance and control.

Crewed aircraft have systems with varying levels of automation that operate without continuous human oversight. Even so, pilots are expected to maintain continuous cognizance and control over the aircraft as a whole. Advanced IA systems could allow unmanned aircraft to operate for extended periods of time without the need for human operators to monitor, supervise, and/or directly intervene in the operation of those systems in real time. This will require that certain critical system functions currently provided by humans, such as "detect and avoid," performance monitoring, subsystem anomaly and failure detection, and contingency decision making, are accomplished by the IA systems during periods when the system operates unattended. Eliminating the need for continuous cognizance and control of unmanned aircraft operations would enable unmanned aircraft to take on new roles that are not practical or cost-effective with continuous oversight. This capability could also improve the safety of crewed operations in situations where risk to a human operator is unacceptably high, workload is too heavy, or the task too monotonous to expect continuous operator vigilance.

Successful development of an unattended operational capability depends on understanding how humans perform their roles in the present system and how these roles are translated to the IA system, particularly for high-risk situations. Eliminating the need for continuous human oversight requires a system architecture that also supports intermittent human cognizance and control.

Specific tasks to be carried out by this research project include the following:

- Investigate human roles, including temporal requirements for supervision, as a function of the mission, capabilities, and limitations of IA systems.
- Develop IA systems that respond safely to the degradation or failure of aircraft systems.
- Develop IA systems to identify and mitigate high-risk situations induced by the mission, the environment, or other elements of the NAS.
- Develop detect-and-avoid IA systems that do not need continuous human oversight.
- Investigate airspace structures that could support UAS operations in confined or pre-approved operating areas using methods such as geofencing.

Modeling and Simulation. Develop the theoretical basis and methodologies for using modeling and simulation to accelerate the development and maturation of advanced IA systems and aircraft.

Modeling and simulation capabilities will play an important role in the development, implementation, and evolution of IA systems in civil aviation because they provide researchers, designers, regulators, and operators with insights into component and system performance without necessarily engendering the expense and risk associated with actual operations. For example, computer simulations may be able to test the performance of some IA systems in literally millions of scenarios in a short time to produce a statistical basis for determining safety risks and establishing the confidence of IA system performance. Researchers and designers are also likely to make use of modeling and simulation capabilities to evaluate design alternatives. Developers of IA systems will be able to train adaptive (i.e., learning) algorithms through repeated operations in simulation. Modeling and simulation capabilities could also be used to train human operators. The committee envisions the creation of a distributed suite of modeling and simulation modules developed by disparate organizations with the ability to be interconnected or networked, as appropriate, based on established standards. The committee believes that monolithic modeling and simulation efforts that are intended to develop capabilities that can "do it all" and answer any and all questions tend to be ineffective due to limitations in access and availability; the higher cost of creating, employing, and maintaining them; the complexity of their application, which constrains their use; and the centralization of development risks. Given the importance of modeling and simulation capabilities to the creation, evaluation, and evolution of IA systems, mechanisms will be needed to ensure that these capabilities perform as intended. A process for accrediting models and simulations will also be required.

Specific tasks to be carried out by this research project include the following:

- Develop theories and methodologies that will enable modeling and simulation to serve as embedded components within adaptive/nondeterministic systems.
- Develop theories and methodologies for using modeling and simulation to coach adaptive IA systems and human operators during training exercises.
- Develop theories and methodologies for using modeling and simulation to create trust and confidence in the performance of IA systems.
- Develop theories and methodologies for using modeling and simulation to assist with accident and incident investigations associated with IA systems.
- Develop theories and methodologies for using modeling and simulation to assess the robustness and resiliency of IA systems to intentional and unintentional cybersecurity vulnerabilities.
- Develop theories and methodologies for using modeling and simulation to perform comparative safety risk analyses of IA systems.
- Create and regularly update standardized interfaces and processes for developing modeling and simulation components for eventual integration.
- Develop standardized modules for common elements of the future system, such as aircraft performance, airspace, environmental circumstances, and human performance.
- Develop standards and methodologies for accrediting IA models and simulations.

Verification, Validation, and Certification. Develop standards and processes for the verification, validation, and certification of IA systems and determine their implications for design.

The high levels of safety achieved in the operation of the NAS largely reflect the formal requirements imposed by the FAA for verification, validation, and certification (VV&C) of hardware and software and the certification of personnel as a condition for entry into the system. These processes have evolved over many decades and represent the cumulative experience of all elements of civil aviation—manufacturers, regulators, pilots, controllers, other operators—in the operation of that system. Although viewed by some as unnecessarily cumbersome and expensive, VV&C processes are critical to the continued safe operation of the NAS. However, extension of these concepts and principles to advanced IA systems is not a simple matter and will require the development of new approaches and tools. Furthermore, the broad range of aircraft sizes, masses, and capabilities envisioned in future civil aviation operations may present opportunities to reassess the current safety and reliability criteria for various components of the aviation system. As was done in the past during the introduction of major new technologies, such as fly-by-wire flight control system and composite materials, the FAA will need to develop technical competency in IA systems and issue guidance material and new regulations to enable safe operation of all classes and types of IA systems.

Specific tasks to be carried out by this research project include the following:

- Characterize and define requirements for intelligent software and systems.
- Improve the fidelity of the VV&C test environment.
- Develop, assess, and propose new certification standards.
- Define new design requirements and methodologies for IA systems.
- Understand the impact that airspace system complexity has on IA system design and on VV&C.
- Develop VV&C methods for products created using nontraditional methodologies and technologies.

ADDITIONAL HIGH-PRIORITY RESEARCH PROJECTS

Nontraditional Methodologies and Technologies. Develop methodologies for accepting technologies not traditionally used in civil aviation (e.g., open-source software and consumer electronic products) in IA systems.

Open-source hardware and software are being widely used in the rapidly evolving universe of IA systems. This is particularly, but not uniquely, true in the active and growing community of hobbyists and prospective entrepreneurs who are developing and operating small unmanned aircraft. Separately, the automotive industry is deploying IA systems using V&V methods different from the methods traditionally used in aviation. The committee believes that there are many potential safety and economic benefits that might be realized in the civil aviation environment by developing suitable methodology that would permit reliable, safe adoption of hardware and software systems of unknown provenance. Although these issues are closely related to issues of V&V and certification, the committee believes they merit independent research attention. This might open up new opportunities for the beneficial deployment of technologies that fall outside traditional uses and applications.

Specific tasks to be carried out by this research project include the following:

- Develop modular architectures and protocols that support the use of open-source products for non-safety-critical applications.
- Develop and mature nontraditional software languages for IA applications.
- Develop paths for migrating open-source, intelligent software to safety-critical applications and unrestricted flight operations.
- Define new operational categories that would enable or accelerate experimentation, flight testing, and deployment of nontraditional technologies.

The final phase of this research project would be accomplished by the VV&C research project as it develops certification standards.

Roles of Personnel and Systems. Determine how the roles of key personnel and systems, as well as related human–machine interfaces, should evolve to enable the operation of IA systems.

Effectively integrating humans and machines in the civil aviation system has been a high priority and sometimes an elusive design challenge for decades. Human–machine integration may become an even greater challenge with the advent of advanced IA systems. Although the reliance on high levels of automation in crewed aircraft has increased the overall levels of system safety, persistent and seemingly intractable issues arise in the context of incidents and accidents. Typically, pilots experience difficulty in developing and maintaining an appropriate mental model of what the automation is doing at any given time. Maintaining an awareness of the operational mode of key automated systems can become especially problematic in dynamic situations. Advanced IA systems will change the specifics of the human performance required by such systems, but it remains to be seen if cognitive requirements for human operators will be more or less stringent. Not only are there significant issues surrounding the proper roles and responsibilities of humans in such systems, but there are also important new questions about the properties and characteristics of the human–machine interface posed by the adaptive/nondeterministic behavior of these systems. The committee believes that in many ways, these are a logical extension of the age-old questions about duties, responsibilities, and skills and training required for pilots, air traffic controllers, and other humans in the system. However, advanced IA systems may permit, or even require, new roles and radical realignment of the more traditional roles of such human actors to achieve some of the benefits envisioned. The importance of these issues cannot be overstated, and, again, realization of projected benefits of autonomy will be constrained by failure to address the issues through research.

Specific tasks to be carried out by this research project include the following:

- Develop human–machine interface tools and methodologies to support operation of advanced IA systems during normal and atypical operations.
- Develop tools and methodologies to ensure effective communication among IA systems and other elements of the NAS.
- Define the rationale and criteria for assigning roles to key personnel and IA systems and assessing their ability to perform those roles under realistic operating conditions.
- Develop intuitive human–machine integration technologies to support real-time decision making, particularly in high-stress, dynamic situations.
- Develop methods and technologies to enable situational awareness that supports the integration of IA systems.

Safety and Efficiency. Determine how IA systems could enhance the safety and efficiency of the civil aviation system.

As with other new technologies, poorly implemented IA systems could put at risk the high levels of efficiency and safety that are the hallmarks of civil aviation, particularly for commercial air transportation. However, done properly, advances in IA systems could enhance both the safety and the efficiency. For example, IA systems have the potential to reduce reaction times in safety-critical situations, especially in circumstances that today are encumbered by the requirement for human-to-human interactions. The ability of IA capabilities to rapidly cue operators or potentially render a fully autonomous response in safety-critical situations could improve both safety and efficiency. IA systems could substantially reduce the frequency of those classes of accidents typically ascribed to operator error. This could be of particular value in the segments of civil aviation, such as general aviation and medical evacuation helicopters, that have much higher accident rates than commercial air transports.

Whether located on board an aircraft or in ATM centers, IA systems also have the potential to reduce manpower requirements, thereby increasing the efficiency of operations and reducing operating costs.

In instances where IA systems make it possible for small unmanned aircraft to replace crewed aircraft, the risks to persons and property on the ground in the event of an accident could be greatly reduced, owing to the reduced damage footprint in those instances, and the risk to air crew is eliminated entirely.

Specific tasks to be carried out by this research project include the following:

- Analyze accident and incident records to determine where IA systems may have prevented or mitigated the severity of specific accidents or classes of accidents.
- Develop and analytically test methodologies to determine how the introduction of IA systems in flight operations, ramp operations by aircraft and ground support equipment, ATM systems, airline operation control centers, and so on might improve safety and efficiency.
- Investigate airspace structures and operating procedures to ensure safe and efficient operations of legacy and IA systems in the NAS.

Stakeholder Trust. Develop processes to engender broad stakeholder trust in IA systems in the civil aviation system.

IA systems can fundamentally change the relationship between people and technology, and one important dimension of that relationship is trust. Although increasingly used as an engineering term in the context of software and security assurance, trust is above all a social term and becomes increasingly relevant to human–technology relationships when complexity thwarts the ability to fully understand a technology's behavior. Trust is not a trait of the system; it is the system status in the mind of human beings based on their perception of and experience with the system. Trust concerns the attitude that a person or technology will help achieve specific goals in a situation characterized by uncertainty and vulnerability.[1] It is the perception of trustworthiness that influences how people respond to a system.

Although closely related to VV&C, trust warrants attention as a distinct research topic because formal certification does not guarantee trust and eventual adoption. Stakeholder trust is also tied to cybersecurity and related issues; trustworthiness depends on the intent of designers and on the degree to which the design prevents both inadvertent and intentional corruption of system data and processes.

Specific tasks to be carried out by this research project include the following:

- Identify the objective attributes of trustworthiness and develop measures of trust that can be tailored to a range of applications, circumstances, and relevant stakeholders.
- Develop a systematic methodology for introducing IA system functionality that matches authority and responsibility with earned levels of trust.
- Determine the way in which trust-related information is communicated.
- Develop approaches for establishing trust in IA systems.

COORDINATION OF RESEARCH AND DEVELOPMENT

All of the research projects described above can and should be addressed by multiple organizations in the federal government, industry, and academia.

The roles of academia and industry would be essentially the same for each research project because of the nature of the role that academia and industry play in the development of new technologies and products.

The FAA would be most directly engaged in the VV&C research project, because certification of civil aviation systems is one of its core functions. However, the subject matters of most of the other research projects are also related to certification directly or indirectly, so the FAA would be ultimately be interested in the progress and results of those other projects as well.

The Department of Defense (DOD) is primarily concerned with military applications of IA systems, though it must also ensure that military aircraft with IA systems that are based in the United States satisfy requirements for operating in the NAS. Its interests and research capabilities coincide with all eight research projects, especially with those on the roles of personnel and systems and operation without continuous human oversight.

[1] J.D. Lee and K.A. See, 2004, Trust in automation: Designing for appropriate reliance, *Human Factors* 46(1): 50-80.

NASA supports basic and applied research in civil aviation technologies, including ATM technologies of interest to the FAA. Its interests and research capabilities also encompass the scope of all eight research projects, particularly modeling and simulation, nontraditional methodologies and technologies, and safety and efficiency.

Each of the high-priority research projects overlaps to some extent with one or more of the other projects, and each would be best addressed by multiple organizations working in concert. There is already some movement in that direction.

The FAA has created the Unmanned Aircraft Systems Integration Office to foster collaboration with a broad spectrum of stakeholders, including DOD, NASA, industry, academia, and technical standards organizations. In 2015, the FAA will establish an air transportation center of excellence for UAS research, engineering, and development. In addition, the NextGen Joint Planning and Development Office (JPDO), which is executing a multiagency research and development plan to improve the NAS, has issued a roadmap for UAS research, development, and demonstration.[2] Efforts such as these are necessary and could be strengthened to assure that the full scope of IA research and development efforts (not just those focused on UAS applications) is effectively coordinated and integrated, with minimal duplication of research and without critical gaps. In particular, more effective coordination among relevant organizations in government, academia, and industry would help execute the recommended research projects more efficiently, in part by allowing lessons learned from the development, test, and operation of IA systems to be continuously applied to ongoing activities.

The recommended research agenda would directly address the technology barriers and the regulation and certification barriers. As noted in Table 4.1, although several research projects would address the social and legal issues, the agenda would not address the full range of these issues. In the absence of any other action, resolution of the legal and social barriers will likely take a long time, as court cases are filed to address various issues in various locales on a case-by-case basis, with intermittent legislative action in reaction to highly publicized court cases, accidents, and the like. A more timely and effective approach for resolving the legal and social barriers could begin with discussions involving the Department of Justice, FAA, National Transportation Safety Board, state attorneys general, public interest legal organizations, and aviation community stakeholders. The discussion of some related issues may also be informed by social science research. Given that the FAA is the federal government's lead agency for establishing and implementing aviation regulations, it is in the best position to take the lead in initiating a collaborative and proactive effort to address legal and social barriers.

CONCLUDING REMARKS

Civil aviation in the United States and elsewhere in the world is on the threshold of profound changes in the way it operates because of the rapid evolution of IA systems. Advanced IA systems will, among other things, be able to operate without direct human supervision or control for extended periods of time and over long distances. As happens with any other rapidly evolving technology, early adapters sometimes get caught up in the excitement of the moment, producing a form of intellectual hyperinflation that greatly exaggerates the promise of things to come and greatly underestimates costs in terms of money, time, and—in many cases—unintended consequences or complications. While there is little doubt that over the long run the potential benefits of IA in civil aviation will indeed be great, there should be equally little doubt that getting there, *while maintaining or improving the safety and efficiency of U.S. civil aviation,* will be no easy matter. Furthermore, given that the potential benefits of advanced IA systems—as well as the unintended consequences—will inevitably accrue to some stakeholders much more than others, the enthusiasm of the latter for fielding such systems could be limited. In any case, overcoming the barriers identified in this report by pursuing the research agenda proposed by the committee is a vital next step, although more work beyond the issues identified here will certainly be needed as the nation ventures into this new era of flight.

[2] JPDO, NextGen UAS Research, Development and Demonstration Roadmap, Version 1.0, March 15, 2012, http://www.jpdo.gov/library/20120315_UAS%20RDandD%20Roadmap.pdf.

1

Autonomous Capabilities and Vision

INTRODUCTION

The development and application of increasingly autonomous (IA) systems for civil aviation (see Boxes 1.1 and 1.2) are proceeding at an accelerating pace, driven by the expectation that such systems will return significant benefits in terms of safety, reliability, efficiency, affordability, and/or previously unattainable mission capabilities. IA systems, characterized by their ability to perform more complex mission-related tasks with substantially less human intervention for more extended periods of time, sometimes at remote distances, are being envisioned for aircraft and for air traffic management (ATM) and other ground-based elements of the national airspace system (NAS) (see Box 1.3). This vision and the associated technological developments have been spurred in large part by the convergence of the substantial investments by the federal government in advanced unmanned aircraft systems (UAS) (see Box 1.4) for military and, to a lesser extent, civil applications and by advances in low-cost, high-capability computing systems; sensor technologies; high-throughput digital communications systems; precise position, navigation, and timing information (e.g., from the Global Positioning System (GPS); open-source hardware and software; and the emergence of an active and prolific hobbyist community that has provided and continues to provide fertile ground for innovation and entrepreneurship that did not exist a decade ago. The burgeoning industrial sector devoted to the design, manufacture, and sales of IA systems is indicative of the perceived economic opportunities that will arise. In short, civil aviation is on the threshold of potentially revolutionary improvements in aviation capabilities and operations associated with IA systems. These systems, however, pose serious unanswered questions about how to safely integrate these revolutionary technological advances into a well-established, safe, and efficiently functioning NAS governed by operating rules that can only be changed after extensive deliberation and consensus.

This report identifies key barriers and suggests major elements of a national research agenda to address those barriers and help realize the benefits that IA systems can make to crewed aircraft, UAS, and ground-based elements of the NAS. This agenda is in large part motivated by the demonstrated capabilities of advanced aerospace vehicles—such as crewed and unmanned military and commercial aircraft, spacecraft, and planetary rovers—that rely on systems that operate with varying levels of autonomy and by the growing desire to have both aircraft and ground-based systems with increased autonomy. Such systems, generally operating within well-defined limits on their ability to act without the direct control of operators, have demonstrated the ability to enable new types of missions, improve safety, and optimize the workload of, for example, pilots in crewed aircraft and remote operators for unmanned aircraft (see Box 1.5). On the other hand, autonomous systems can introduce uncertainties if

BOX 1.1
Civil Aviation

In this report, "civil aviation" is used to refer to all nonmilitary aircraft operations in U.S. civil airspace. This includes operations of civil aircraft as well as nonmilitary public use aircraft (that is, aircraft owned or operated by federal, state, and local government agencies other than the Department of Defense). In addition, many of the IA technologies that would be developed by the recommended research projects would generally be applicable to military crewed and/or unmanned aircraft for military operations and/or other operations in the NAS.

BOX 1.2
Increasingly Autonomous Systems

A fully autonomous aircraft would not require a pilot; it would be able to operate independently within civil airspace, interacting with air traffic controllers and other pilots just as if a human pilot were on board and in command. Similarly, a fully autonomous ATM system would not require human air traffic controllers. This study is not focused on these extremes (although it does sometimes address the needs or qualities of fully autonomous unmanned aircraft). Rather, the report primarily addresses what the committee calls "increasingly autonomous" (IA) systems, which lie along the spectrum of system capabilities that begin with the abilities of current automatic systems, such as autopilots and remotely piloted (nonautonomous) unmanned aircraft, and progress toward the highly sophisticated systems that would be needed to enable the extreme cases. Some IA systems, particularly adaptive/nondeterministic IA systems, lie farther along this spectrum than others, and in this report such systems are typically described as "advanced IA systems."

BOX 1.3
National Airspace System

The NAS is "the common network of U.S. airspace; air navigation facilities, equipment, and services; airports or landing areas; aeronautical charts, information and services; rules, regulations, and procedures; technical information; and manpower and material" (Integration of Civil Unmanned Aircraft Systems [UAS] in the National Airspace System [NAS] Roadmap, FAA, 2013). Some NAS facilities are jointly operated by the FAA and the Department of Defense. IA systems could be incorporated into airport ground systems such as snow plows. However, the greatest technological, social, and legal challenges to the use of IA systems in civil aviation are associated with their use in aircraft and air traffic management systems, and the report does not specifically address the use of IA systems in airport ground systems.

BOX 1.4
Unmanned Aircraft/Crewed Aircraft

An unmanned aircraft is "a device used or intended to be used for flight in the air that has no onboard pilot. This device excludes missiles, weapons, or exploding warheads, but includes all classes of airplanes, helicopters, airships, and powered-lift aircraft without an onboard pilot. Unmanned aircraft do not include traditional balloons (see 14 CFR Part 101), rockets, tethered aircraft and un-powered gliders." A UAS is "an unmanned aircraft and its associated elements related to safe operations, which may include control stations (ground-, ship-, or air-based), control links, support equipment, payloads, flight termination systems, and launch/recovery equipment" (Integration of Civil Unmanned Aircraft Systems [UAS] in the National Airspace System [NAS] Roadmap, FAA, 2013). UAS include the data links and other communications systems used to connect the UAS control station, unmanned aircraft, and other elements of the NAS, such as ATM systems and human operators. Unless otherwise specified, UAS are assumed to have no humans on board either as flight crew or as passengers. "Crewed aircraft" is used to denote manned aircraft; unless specifically noted otherwise, manned aircraft are considered to have a pilot on board.

BOX 1.5
Operators

In this report, the term "operator" generally refers to pilots, air traffic controllers, airline flight operations staff, and other personnel who interact directly with IA civil aviation systems. "Pilot" is used when referring specifically to the operator of a crewed aircraft. With regard to unmanned aircraft, the FAA says that "in addition to the crewmembers identified in 14 CFR Part 1 [pilots, flight engineers, and flight navigators], a UAS flight crew includes pilots, sensor/payload operators, and visual observers, but may include other persons as appropriate or required to ensure safe operation of the aircraft" (Integration of Civil Unmanned Aircraft Systems [UAS] in the National Airspace System [NAS] Roadmap, FAA, 2013). Given that the makeup, certification requirements, and roles of UAS flight crews are likely to evolve as UAS acquire advanced IA capabilities, this report refers generally to UAS operators as the flight crew rather than specifically as pilots.

they are not thoroughly assessed over a broad range of normal and abnormal operating conditions. Autonomous systems may reduce operational costs by reducing the need for highly trained human operators, but these savings may be partially or wholly offset by the need for substantive advances in system design, testing, and certification processes. As a result, it remains to be seen what benefits IA systems in various civil aviation applications will be able to provide without degrading safety, reliability, mission performance, and/or net costs. This is especially important in commercial aviation, where extremely high levels of safety and a very competitive business climate are the norm.

CHARACTERISTICS AND FUNCTION OF IA SYSTEMS

At the highest level, autonomy implies the ability of the system (often a machine) to perform tasks that involve dynamically executing a "decision cycle" in much the same fashion as a human. That decision cycle can

FIGURE 1.1 The classic OODA loop.

be described in a number of ways; perhaps the simplest model is the so-called OODA loop[1] (see Figure 1.1). The acronym stands for the four-step process used to execute virtually any task:

- Observe,
- Orient,
- Decide, and
- Act.

The concept was originally developed during the Vietnam era to describe and advocate effective air combat tactics. An autonomous system, be it human or machine, first observes by sensing or acquiring information about the environment from other relevant sources. It then orients itself toward the task at hand. This second step infers a number of functions that can encompass information fusion, contextual interpretation, the integration of learned behaviors, and even inferences about future events. In the robotics community, a number of the capabilities associated with this step are often referred to in aggregate as perception. The third step then involves making a decision based on the task objectives and the results of the prior steps. This requires that the system be capable of implementing an appropriate action that accomplishes the task. Once the action is complete, the cycle repeats as the system observes the consequences of the action as well as changes in the environment caused by other factors. The OODA loop has found application in many diverse areas, including business and law.[2,3]

Within the aviation context, this decision cycle may be implemented one time in isolation (e.g., a simple aircraft heading change) or repeated with regularity (e.g., a digital flight control system operating at 30 decision cycles per second), depending on the task. The OODA concept of autonomy applies to tasks that range from lower-level functions, such as stabilization and basic maneuvering of an aircraft (e.g., a fly-by-wire control system), to high-level mission decisions and even to the accomplishment of a complete mission. Some of these capabilities exist today, while others will be possible only as IA technologies mature over time.

There is a continuum that describes the roles of humans and machines when they interact. This continuum ranges from systems at one end of the spectrum, where the human is in total control, to systems at the other end, where they operate without the need for human interaction. Some systems operate at the extremes. For example, antilock braking systems and airbag systems on cars are fully automatic, deciding on their own when to act. However, fully automatic systems as well as fully autonomous systems depend on humans to define and limit the scope of their authority and the range of possible actions. Movement along the continuum typically does not eliminate or

[1] Frans P.B. Osinga, 2007, *Science, Strategy and War: The Strategic Theory of John Boyd (Strategy and History)*, London and New York: Routledge, Taylor and Francis Group.

[2] Chet Richards, 2004, *Certain to Win: The Strategy of John Boyd, Applied to Business,* Bloomington, Ind.: Xlibris Corporation.

[3] A.S. Dreier, 2012, *Strategy, Planning & Litigating to Win: Orchestrating Trial Outcomes with Systems Theory, Psychology, Military Science and Utility Theory*, Boston, Mass.: Conatus Press.

TABLE 1.1 Characteristics of Advanced Automation and Autonomy

Characteristics	Advanced Automation	Advanced Autonomy
Augments human decision makers	Usually	Usually
Proxy for human actions or decisions	Usually	Usually
Reacts at cyber speed	Usually	Usually
Reacts to the environment	Usually	Usually
Reduces tedious tasks	Usually	Usually
Robust to incomplete or missing data	Usually	Usually
Adapts behavior to feedback (learns)	Sometimes	Usually
Exhibits emergent behavior	Sometimes	Usually
Reduces cognitive workload for humans	Sometimes	Usually
Responds differently to identical inputs	Sometimes	Usually
Addresses situations beyond the routine	Rarely	Usually
Replaces human decision makers	Rarely	Potentially
Robust to unanticipated situations	Limited	Usually
Adapts behavior to unforeseen environmental changes	Rarely	Potentially
Behavior is determined by experience rather than by design	Never	Usually
Makes value judgments (weighted decisions)	Never	Usually
Makes mistakes in perception and judgment	N/A	Potentially

diminish the importance of the humans to the operations of the system, but it does change their role. In fact, the human's role becomes more rather than less important when moving toward the autonomous end of the spectrum because it is so important to assure that the systems are properly designed, tested, deployed, and monitored to ensure the aircraft's continued airworthiness.

The International Civil Aviation Organization (ICAO) is an agency of the United Nations that develops aviation standards that member states may use when developing national aviation regulations. ICAO has defined an autonomous aircraft as "an unmanned aircraft that does not allow pilot intervention in the management of the flight."[4] This is an unfortunate definition since it is unlikely that any unmanned aircraft would be designed to "not allow" human intervention. On the contrary, both civil and military unmanned aircraft are designed to allow operators to intervene whenever they are in contact with their aircraft. In addition, autonomous aircraft would not necessarily be unmanned, in that properly equipped aircraft would be able to operate autonomously even with a crew and/or passengers on board. Accordingly, it would be more accurate and more useful to define an autonomous aircraft as "an aircraft that does not *require* pilot intervention in the management of the flight." This slightly altered definition suggests a policy that focuses on assuring the safety of unmanned aircraft operations without the presumption that safety will necessarily require continuous oversight by an onboard or remote pilot. However, even this definition fails to describe the full range of capabilities that make systems autonomous, nor does it provide an understanding of how autonomous or IA systems can be distinguished from automated systems, which are quite common in the NAS. The differences between automated and autonomous systems—and the characteristics shared by both—are addressed below.

Automated systems can be very complex and behave in a way that some might say reflects a form of artificial intelligence and, thus, embodies a degree of autonomy. The difference may be in the eye of the beholder. In fact, complex automatic or IA systems share many of the same operational, technical, and policy challenges.

Table 1.1 describes the characteristics of systems with advanced automation or autonomy. There are no absolutes and no specific characteristics that label a system as being automatic or autonomous. Thus, this report focuses on barriers and required research necessary to make relevant technological advances and less on the definitions. Also, because there are no absolutes associated with a formal definition of an autonomous system, this report uses the term *increasingly autonomous* as a label to discuss the operational, technical, and policy barriers to implementation in civil aviation and in presenting a research agenda.

[4] International Civil Aviation Organization, 2011, Circular 328 AN/190—Unmanned Aircraft Systems (UAS).

TRENDS IN THE USE OF AUTONOMY IN AVIATION

Early Aviation Years

Automation and autonomy have been used in aviation almost from the beginning. The Sperry Autopilot developed in 1912 allowed straight and level flight without the pilot operating the stick. This system was demonstrated in 1914 at the Paris Concours de la Sécurité en Aéroplane show, with an aircraft that performed a stable fly-by with both pilots standing on the wings. This autopilot system was incorporated into the first unmanned aircraft, the Kettering Bug, in 1918.

The Second World War saw the expansion in the use of flight control systems. Navigation systems were coupled with autopilots to facilitate flying in adverse weather conditions and at night. In 1947 a C-53 conducted the first transatlantic flight with the aircraft completely under the control of the autopilot, including takeoff and landing. The beginnings of fly-by-wire technology were seen in the late 1950s and 1960s. These systems used an analog computer as an electrical control system, with a mechanical backup control system. The Avro Canada CF-105 Arrow was the first aircraft to implement such an approach, in 1958, and it was also used on the Concorde (1969) and the Apollo Lunar Landing Research Vehicle (1964), although the latter did not include a mechanical backup. The first production aircraft to use an analog fly-by-wire control system without a mechanical backup was the F-16A/B (1976). The F-16 was also designed for relaxed stability, which required an analog computer to maintain stability and keep it in the air; a quadruply redundant configuration was used for reliability.

Early radio-controlled unmanned aircraft (e.g., the Radioplane RP-4/OQ-1) appeared in the 1930s and 1940s. This type of aircraft, which was controlled by pilots on the ground, soon gave way to target drones (e.g., Ryan Firebee), which had greater endurance and range and were no longer remotely piloted. Instead, they used simple dead-reckoning navigation.

The Vietnam War saw the evolution of target drones into UAS performing intelligence, surveillance, and reconnaissance missions as well as electronic warfare missions (e.g., Ryan AQM-34 and AQM-91A). These systems had long-range navigation capability, autonomous pre-programmed missions, and the ability to react to internal monitoring. As the Vietnam era ended, the United States placed less emphasis on UAS development, while Israel began to ramp up its efforts (e.g., Tadiran Mastiff).

Introduction of Digital Technology

The replacement of analog computers by digital computers provided the necessary foundation for significant autonomy enhancements in crewed and unmanned aircraft. After developing a digital fly-by-wire control system for the Apollo command module and lunar lander, the National Aeronautics and Space Administration (NASA) validated the concept of an all-digital, fly-by-wire flight control system in 1972 using an F-8C Crusader with no mechanical backup. This would become the forerunner to the all-digital fly-by-wire control system used by the Space Shuttle Orbiter Enterprise (1977) for unpowered approach and landing tests and by the upgraded F-16C/D (1984). The Airbus A320 (1987) was the first commercial transport to include a full-authority digital fly-by-wire flight control system, which allowed it to be equipped with flight envelope protection. Virtually all large commercial transports, including business jets, now incorporate digital fly-by-wire flight control systems.

U.S. interest in military UAS ramped up in the mid-1980s with several programs conducted by the Defense Advanced Research Projects Agency that were focused on long-endurance operations, as well as the Navy's early experiments with Pioneer UAS. In the early 1990s, during the first Gulf War, UAS such as the Predator A were acquired in large numbers. These systems rapidly gained the respect of the operational community for their ability to provide essentially uninterrupted, continuous surveillance of targets and other points of interest using another new technology, streaming video. During that period a number of enhancements were introduced to the largest of these systems, including satellite communications connectivity and basic mission-level automation. Waypoint navigation enabled unmanned aircraft to fly complex flight paths without the aid of real-time pilot interaction. The combination of these technologies permitted ground operators to be stationed at long distances from the aircraft in flight. Communications latency, however, precluded remote operators from piloting aircraft in real time, in terms of controlling aircraft state and dynamics. In addition, automatic takeoff and landing improved safety and reliability by eliminating the need for real-time interactions between ground operators and unmanned aircraft during these time-critical events. The collective increase in capability provided by automated systems, along with other mission risk mitigation software, led to the elimination of many traditional cockpit controls. For example, stick-and-rudder pedal controls were not included in the design of the first ground station for Global Hawk UAS (1998) and subsequent advanced UAS.

Modern Age

Precision GPS for both military and commercial applications was introduced in the late 1990s. This capability greatly enhanced autonomous navigation. The merger between precision digital control and precision navigation allowed for an expansion of the mission envelope by UAS. Intelligence, surveillance, and reconnaissance capabilities were improved to the point that UAS were providing more information than ground personnel and systems could process in a timely fashion. Beginning in 2001, UAS, notably the Predator MQ-1A, were equipped with weapons, and after completion of testing they began to perform integrated surveillance and strike missions.

After 9/11 the number of military UAS skyrocketed. Fixed and rotary-wing unmanned aircraft as small as a few inches (micro air vehicles) and as large as a commercial transport have found their way into a variety of applications. The operational envelope has also been expanded to include high subsonic speeds, as exhibited by the X-47B unmanned combat air system. The array of missions has expanded to include air-to-ground combat and rapid resupply transport. Advanced features comprise increasingly sophisticated autonomous capabilities, including contingency management so that the aircraft can continue operations even if, for example, communications are lost or aircraft flight characteristics are degraded by damage. Because of the urgent demand for UAS in theaters of conflict and the military's less stringent requirements for assuring operational safety of UAS prior to use in military airspace, UAS have served as pioneers in the development and test of IA capabilities.

Industry and government are using UAS in some other countries. For example, the Yamaha RMAX small unmanned helicopter is widely used in Japan for precision agriculture applications, including crop dusting. In the United States, the government is using UAS for civil missions such as border patrol, weather monitoring, and disaster response. The use of UAS by local law enforcement is still in its infancy. The deployment of UAS for commercial applications in the United States has been extremely limited but is likely to expand rapidly as technical and regulatory barriers are addressed. The first routine commercial operations in the United States are likely to involve small unmanned aircraft operating in accordance with the standards currently established for the operation of model aircraft, which are as follows:[5]

> a. Select an operating site that is of sufficient distance from populated areas. The selected site should be away from noise-sensitive areas such as parks, schools, hospitals, churches, etc.
>
> b. Do not operate model aircraft in the presence of spectators until the aircraft is successfully flight tested and proven airworthy.
>
> c. Do not fly model aircraft higher than 400 feet above the surface. When flying aircraft within 3 miles of an airport, notify the airport operator, or when an air traffic facility is located at the airport, notify the control tower, or flight service station.
>
> d. Give right of way to, and avoid flying in the proximity of, full-scale aircraft. Use observers to help if possible.
>
> e. Do not hesitate to ask for assistance from any airport traffic control tower or flight service station concerning compliance with these standards.

VISIONS FOR INCREASED AUTONOMY IN CIVIL AVIATION

The committee's overarching vision for increased autonomy in civil aviation is that all aircraft and ground systems will be imbued with new or improved capabilities that enable them to function more safely, reliably,

[5] FAA, 1981, Advisory Circular 91-57, Model Aircraft Operating Standards, June 9, http://www.faa.gov/documentLibrary/media/Advisory_Circular/91-57.pdf.

and efficiently over an expanded array of missions constrained only by technological limitations and acceptable margins of risk and cost. A mix of crewed and unmanned aircraft will operate in shared airspace, guided by ATM systems with distributed responsibilities and authorities to assure safe separation and maximize traffic flow during normal and abnormal operating conditions (e.g., in adverse weather). In addition, both air and ground systems are designed to minimize the susceptibility of individual systems and the NAS as a whole to various failure modes.

Detailed information on the potential benefits of achieving the visions for crewed aircraft, unmanned aircraft, and ATM appears in Chapter 2.

Crewed Aircraft

The committee's vision for increased autonomy in crewed aircraft is that (1) flight crews will be better able to perform existing, proposed, and potential tasks because onboard IA systems will provide guidance and aid for the requisite tasks, monitor the performance of both the pilot and the system for predictors of failure, and—within the limits of authority granted to particular IA systems—intervene when necessary to assure flight safety; (2) single-pilot operations will be a viable option for air carriers because the onboard IA systems will perform functions that are currently being performed by humans; and (3) the need to have a pilot on board passenger aircraft may be eliminated.

Unmanned Aircraft

The committee's vision for increased autonomy in unmanned aircraft is that this class of vehicle takes on a growing array of missions as advanced IA systems improve the ability of unmanned aircraft to self-separate, to autonomously coordinate their missions, and to operate unattended for extended periods of time in segregated airspace and, ultimately, in airspace shared by crewed aircraft, while also reducing the cost and increasing the flexibility of UAS operations.

Air Traffic Management

The committee's vision for increased autonomy in ATM is that air traffic controllers, traffic flow managers, and other personnel will be better able to perform existing, proposed, and potential tasks because the IA systems that they use will provide guidance and aid for the requisite tasks, monitor their performance and that of the entire system for predictors of failure, and provide functionalities that heretofore could only be provided by humans, thereby increasing the efficiency of operations in the NAS.

2

Potential Benefits and Uses of Increased Autonomy

This chapter begins with a discussion of the potential benefits that increasingly autonomous (IA) systems may be able to provide to civil aviation, in terms of general criteria such as safety and cost. It then addresses specific applications of IA systems for air traffic management (ATM) and for various classes of aircraft. This chapter also reviews the use of IA systems for nonaviation applications.

POTENTIAL BENEFITS OF INCREASED AUTONOMY FOR CIVIL AVIATION

Estimates of the current market for commercial UAS range from $5.9 billion annually in the United States to $89 billion worldwide.[1,2] The drivers behind the growth of UAS include the potential to increase safety and reliability, reduce costs, and enable new operational capabilities. However, unless IA systems are implemented in a careful and deliberate manner, the actual benefit of IA systems could be limited or they could even reduce safety and reliability or increase costs. In addition, the potential benefits that accrue from the introduction of advanced IA systems in civil aviation, the associated costs, and the unintended consequences that are likely to ensue will not fall on all stakeholders equally. In fact, some stakeholders may not benefit at all from advances that greatly benefit other existing or new stakeholders. For example, making it easier for farmers to operate unmanned aircraft over their fields may be of great economic benefit to them, but it may inconvenience general aviation pilots if they must adjust their operations to avoid conflicts with agricultural unmanned aircraft systems (UAS). In addition, it would take considerable resources for the Federal Aviation Administration (FAA) to address issues related to the certification and operation of IA systems on crewed and unmanned aircraft.

Safety and Reliability

Commercial air transportation in the United States operates at unprecedented levels of safety and reliability. The fatal accident rate for U.S. commercial air carrier operations is so close to zero (literally zero fatalities for

[1] Nick Wingfield and Somini Sengupta, Drones set sights on U.S. skies, *New York Times,* February 17, 2012, http://www.nytimes.com/2012/02/18/technology/drones-with-an-eye-on-the-public-cleared-to-fly.html.

[2] Daisy Carrington and Jenny Soffel, 15 ways drones will change your life, *CNN.com*, November 18, 2013, http://edition.cnn.com/2013/11/03/business/meet-your-friendly-neighborhood-drones/.

2010-2012[3]) that further increases in safety and reliability can only be made in very small increments. Accident rates in some other segments of civil aviation, however, are substantially higher. For example, general aviation accidents resulted in 432 fatalities in 2012,[4] most of them attributed to loss of control, controlled flight into terrain, and other pilot-induced errors. Most general aviation flights have a single pilot on board, and advanced IA systems could carry out some of the duties of a copilot, such as monitoring systems and identifying potential solutions to emerging problems. However, IA systems must be relatively inexpensive if they are to improve general aviation safety. Many existing safety systems, such as the Traffic Collision Avoidance System, that are required on commercial air transports are installed on very few general aviation aircraft because of their cost.

Neither the general public nor the aviation community would tolerate any decrement in safety as a result of changes in the NAS associated with introduction of advanced IA systems. Fortunately, IA systems have the potential to increase safety because automated and autonomous systems do not become distracted, do not tire, are not affected by emotions such as fear, and will not be influenced by other outside pressures, such as the perceived need to complete a flight to some specific airport at a given time. Further, advanced IA systems can have the potential to accommodate a wide variety of dynamic conditions and to react at cyberspeeds to abrupt changes. Theoretically, properly designed IA systems might compensate for certain limitations in human performance that are associated with incidents and accidents.

In general, safety and reliability would be enhanced by IA technologies that are able to execute tasks such as the following:

- Adapt to changing patterns and preferences;
- Remain vigilant at all times;
- Increase the ability to tailor actions to specific circumstances and add flexibility to plans so that they better fit the immediate demands of the situation;
- Increase the situational awareness of human operators by presenting information in a context-sensitive fashion;
- Monitor human actions and alert and/or intervene to prevent errors from causing incidents or accidents; and
- React quickly to avoid critical situations such as a collision.

In addition, advanced IA capabilities that enable UAS to replace crewed aircraft eliminate the need for an air crew and, thus, eliminate the possibility of onboard fatalities in case of an accident. If an aircraft crashes into a populated area, casualties on the ground are likely to be far less if the accident involves a small unmanned aircraft instead of a crewed aircraft, which is certain to be much larger and carry more fuel. To fully realize these safety improvements, safety standards established for the certification and operation of UAS will need to take into account how the accident rate of UAS compares to the accident rate of the crewed aircraft that UAS would replace and of the potentially large number of UAS that would be used for applications that are not currently performed using aircraft.

Costs

IA systems have the potential to reduce costs by reducing the need for highly skilled operators, by enabling more efficient operations, and by relying on UAS to conduct missions that would otherwise be executed by crewed aircraft.

The roles of humans in the NAS will change as advanced IA systems are introduced. The need for humans to fill some existing roles may be reduced or disappear altogether, while other roles may become more important or more prevalent. In addition, some new roles may arise. Coordination among all human operators will remain essential, though such coordination will likely be assisted by and necessarily include coordination with IA systems that have taken on important roles. The changing roles of human operators will reduce costs to the extent the NAS can continue to operate with fewer highly skilled operators. In some cases, highly skilled operators such as pilots could be replaced by less-skilled personnel on the ground (UAS operators). For example, remote crew might be used to augment the onboard crew to reduce crew requirements during long-haul flights and to provide better

[3] National Transportation Safety Board, 2012, "Review of Accident Data," http://www.ntsb.gov/data/ aviation_stats.html.

[4] Ibid.

opportunities for crew rest.[5] However, IA systems could also increase the need for highly skilled support staff, in particular the engineers needed to develop IA systems, develop new verification, validation, and certification (VV&C) standards, and assure that they perform safely and reliably over their lifetime.

The traditional approach to dealing with situations where the cognitive demand required to complete a given set of specified tasks is beyond the cognitive capacity of a single person is to divide the tasks among multiple personnel so that no individual is overloaded. For example, in airspace with a large number of aircraft, air traffic control tasks may be divided so that aircraft in a particular airspace volume are under the control of an R-side (or radar) controller, who is responsible for monitoring the position of aircraft and exercising tactical control, and a D-side (or data) controller, who is responsible for managing the data required to make and communicate decisions while also exercising strategic control. Increased autonomy in ATM could lead to the shedding of certain tasks that are currently performed by humans. This in turn could increase the number of aircraft that are handled by each pair of R- and D-side controllers and/or (in the extreme) eliminate the need for one of the controllers. These changes could increase the efficiency with which aircraft are handled and, thus, the efficiency of NAS operations while also reducing the need to increase the number of highly skilled controllers as the level of air traffic increases.

The cost of developing, manufacturing, and operating unmanned aircraft—especially small unmanned aircraft—can be much less than the cost of crewed aircraft that would execute the same missions. Thus, in many cases costs could be reduced by replacing crewed aircraft with highly capable UAS. In other cases, UAS will enable new economic activity by enabling businesses to provide services that could not be profitably executed with crewed aircraft.

UAS Operational Capabilities

IA systems could enhance existing UAS capabilities and enable new capabilities. For example, IA systems applied to UAS have the potential to do all of the following:

- Increase precision in agriculture by tailoring actions to specific situations, as in adjusting fertilizer or water applications within a particular field.
- Add flexibility to mission plans so that they better fit the situation and context, as in policing.
- Provide long-duration surveillance to better monitor critical areas, as in tracking chemical spills through a watershed.
- Take on dangerous missions currently executed by crewed aircraft, as in fighting forest fires.
- Facilitate coordination between different groups responsible for different aspects of a situation, as in urban firefighting teamwork.
- Learn from experience to adapt to patterns and preferences, as in fixed-obstacle avoidance during unmanned rotorcraft urban transit.
- Respond rapidly to sudden critical events, as in collision avoidance.
- Investigate from a distance, allowing experts to view local conditions, as in structural engineering inspections of damaged buildings following an earthquake.
- Act at a distance, allowing people to stand off from hazardous conditions and yet act to assess and respond to needs, as in response to industrial disasters with chemical, biological, structural, or radiological risks.
- Investigate from within, as in building interior inspections.
- Manage communication assets to maximize reliable connectivity and bandwidth utilization, as in replacing cell towers damaged in natural disasters.
- Make it easier and more affordable to observe and inspect large areas, as in the inspection of long-distance pipelines.
- Improve safety, as in general aviation, including light sport aircraft and amateur-built aircraft.
- Enable operations beyond the line of sight (that is, without continuous communication with ground personnel), as in monitoring the environment in remote locations.

[5] D. Learmount, 2013/2014, When will we drop the pilot?, *Flight International*, December-January.

- Increase safety in off-nominal flight conditions by exploiting precision and rapid response characteristics to expand an aircraft's operational envelope, as in rotorcraft operating in congested airspace.

The design space for unmanned aircraft is much larger than that for crewed aircraft. The absence of human beings removes many limitations associated with human frailties, from physiological limitations (e.g., acceleration limits, motion sickness, and other disorientation susceptibilities) to environmental adaptivity (e.g., pressurization and oxygen content) and more. It opens up a number of aircraft design options that can expand the operational envelope in a number of dimensions.[6]

- *Long endurance.* This capability has already been realized in a number of current operational systems that are primarily used for surveillance. The ability of a platform to stay aloft for periods that far exceed typical crew endurance has been demonstrated in a number of unmanned aircraft.
- *Size/scale.* This trend is also already under way, exploiting aircraft that are physically diminutive in comparison to even the smallest human pilot. Small unmanned aircraft continue to gain prominence and increased presence in combat applications and are being considered for civil and commercial applications ranging from law enforcement to pizza delivery.
- *Maneuvering performance and agility.* Unmanned platforms are inherently capable of sustaining accelerations well beyond the tolerance of the most fit and adaptable human pilots, improving maneuverability to unprecedented levels.
- *Unique configuration options.* There is no need to accommodate the constraints on size, volume, or shape imposed by an onboard human presence, increasing the possibilities for innovative airframe shapes and configurations.
- *In-flight orientation.* Vehicle orientation in flight can be completely arbitrary at any time during the flight profile, limited only by mission needs and considerations. Novel exploitation of this attribute, coupled with the increased configuration options mentioned above, can lead to innovative concepts that capture the best of fixed- and rotary-wing designs in a single configuration.
- *Design lifetime.* Small unmanned aircraft can be so inexpensive that they become economical even if they are designed to have a relatively short lifetime. A short lifetime is untenable for crewed aircraft, but it becomes a practical alternative for a small unmanned vehicle that could serve in a number of new applications and remake the conventional life-cycle cost paradigm.
- *Novel launch and recovery methods.* With IA systems, options for unconventional launch and recovery methods increase dramatically. These concepts would add new options for achieving runway independence and reduce infrastructure costs for a range of mission applications.

USES OF INCREASED AUTONOMY IN CIVIL AVIATION

Air Traffic Management

Current ATM procedures vary little from those followed by air traffic controllers 50 years ago. While automation has increased the availability of information and facilitates communications among controllers, decisions are still made by human beings. Some automatic alerting functions, such as conflict alerts, are designed to warn controllers when situations arise that need their attention, and some decision support tools assist controllers in determining the suitability of potential courses of action. Using the Observe, Orient, Decide, and Act (OODA) loop as a template, IA systems might be used in air traffic management as follows:

- *Observe.* Scan the environment and gather data from it. IA systems could assist controllers by monitoring many more data sources than humans can effectively monitor.

[6] M.S. Francis, 2011, Unmanned air systems—Challenge and opportunity, *Journal of Aircraft* 49(6): 1652-1665.

- *Orient*. Synthesize the data into information. IA systems could monitor voice and data communications for inconsistencies and mistakes; monitor aircraft tracks for deviations from clearances; identify flight path conflicts; monitor weather for potential hazards as well as potential degradations in capacity; and detect imbalances between airspace demand and capacity.
- *Decide*. Identify options and determine a course of action. IA systems could identify and evaluate traffic management options for controllers and flow managers and recommend a particular course of action.
- *Act*. Follow through on decisions. IA systems could use voice and/or datalink communications to relay controller decisions to pilots, traffic flow managers, and other controllers. This could be particularly beneficial given that controllers spend so much of their time communicating.

The effectiveness of IA systems for ATM functions would be enhanced by the inclusion of compatible airborne IA systems that facilitate the exchange of information between pilots and aircraft and controllers and ATM systems.

In addition, IA systems could be used as an interactive training tool for initial and recurrent training of air traffic controllers. This could reduce the cost and duration of controller training and increase its effectiveness.

Fixed-Wing Transport Aircraft

Federal Aviation Regulations (that is, Title 14 of the Code of Federal Regulations) direct pilots to remain "well clear" of other aircraft that have the right of way. Self-separation of aircraft is achieved through each pilot's responsibility to see and then avoid conflicting traffic by maneuvering so as to remain well clear. Unmanned aircraft will require detect-and-avoid capabilities to meet self-separation requirements. In addition, regulatory changes will be needed to permit IA detect-and-avoid systems to satisfy the intent of current see-and-avoid regulatory requirements. However, the use of detect-and-avoid systems need not be limited to UAS. Such systems could also be incorporated into crewed aircraft to enhance their ability to self-separate.

Well-trained and qualified pilots are a source of resilience that contributes to the ultrahigh safety and the efficiency of the commercial transport operations. Pilots provide resilience to mitigate the safety and operational risks posed by weather, traffic congestion, workload, human errors, and system malfunctions. These risks include the following:

- Changes in operational circumstances, such as revised air traffic clearances due to weather and traffic congestion.
- Errors made by pilots, dispatchers, maintainers, air traffic personnel, or equipment designers.
- Equipment limitations and lack of availability.
- Equipment malfunctions.

Appropriately designed IA systems with carefully considered human interfaces could mitigate each of these risks.

Rotorcraft

Rotary-wing aircraft are extensively used throughout the world in a variety of civil aviation roles, many of which cannot be performed by fixed-wing aircraft. Rotary-wing aircraft missions include short-haul commercial passenger and cargo transportation, disaster relief, search and rescue, medical evacuation, construction, logging, fire fighting, pipeline management and inspection, drug interdiction, border control, traffic management, law enforcement, and agricultural services.[7] Rotorcraft are particularly important for operations to destinations such as oil platforms, hospitals, and accident scenes without runways for fixed-wing aircraft.

The role of the rotorcraft pilot has been continuously influenced by the instantiation of IA systems. These systems have significantly reduced pilot workload and increased operational effectiveness and safety. Systems of particular interest include automatic turbine engine/rotor speed governing systems, stability augmentation sys-

[7] *Sikorsky Archives News*, January 2014, Igor I. Sikorsky Historical Archives, Inc., Stratford, Conn., http://www.sikorskyarchives.com/pdf/news%202014/News%20Jan2014sm.pdf.

tems, automatic flight control systems, velocity and altitude hold systems, station-keeping systems, and autopilot and autonavigation systems. The accelerated use of IA systems can be expected to greatly facilitate the role of rotorcraft in the coming decades. In the process, the role of the human operator may be transformed from pilot to aircraft manager.

Optionally piloted rotorcraft are already under development for civil and military applications. Such aircraft could allow pilots to act as medical technicians or rescue crew in an emergency and to manage both their own aircraft and companion UAS. Optionally piloted rotorcraft would also reduce crew size, which would increase payload capacity and lower crew and training costs.

General Aviation

General aviation operations encompass a wide variety of equipage, pilots, and missions. Many older general aviation aircraft are equipped only for operations under visual flight rules (VFR), while many newer aircraft have redundant autopilot and flight management systems with glass cockpit displays. Pilots range in skill level from student pilots to commercial pilots, flight instructors, and air show performers with thousands of flight hours. As already noted, the accident rate for general aviation is much higher than that for commercial air carriers. Over the past decade, about 70 percent of the fixed-wing general aviation accidents were attributed to pilot-related causes, mostly associated with flight planning and decision making, particularly in high-workload situations associated with takeoff, approach, low-altitude maneuvering, and adverse weather.

IA systems have the potential to appreciably improve general aviation safety. Pilot-assist systems can facilitate in-flight decision making. Advanced autopilots can reduce the potential for loss of control through a combination of warnings and recovery options analogous to stability augmentation systems found in cars. Depending upon the design of the system and the nature of the emergency, IA systems could be designed to help a pilot avoid or recover from a loss-of-control situation either with or without direct input from the pilot. IA systems with navigation, detect-and-avoid, and flight planning and guidance capabilities would have the potential to eliminate the majority of general aviation accidents occurring today.

IA systems for general aviation would need to be inexpensive to support widespread adoption. However, apps currently available for tablets and smartphones offer better situational awareness than legacy general aviation avionics. Such tools are increasingly migrating into the cockpit as real-time situational awareness and decision aids. Autonomy research for civil aviation can improve the capabilities and quality of such tools as well as the process by which they are tested and certified.

Unmanned Aircraft Systems

UAS that currently exist or that could easily be developed can execute a wide variety of missions. These missions make use of a wide range of unmanned aircraft types, from diminutive micro air vehicles to large, high-speed aircraft the size of commercial transports. Some missions would be executed by fixed-wing unmanned aircraft, while others would be best served by unmanned rotorcraft with one, two, four, six, or eight rotors[8] or by airships of various sizes.

Small, low-cost unmanned aircraft could inexpensively execute a wide variety of local observation and inspection missions, such as traffic and crop monitoring. Larger unmanned aircraft could execute missions with requirements for long endurance and/or large area coverage, such as inspection of oil or natural gas pipelines, wildlife observation, or communications relay. Large unmanned aircraft could also execute missions involving heavy payloads, such as advertising, crop dusting, construction, and fire fighting. Unmanned aircraft of various sizes could be flown by the general public for general aviation and personal recreation and by industry, academia, and the government for research, development, test, and certification.

The United States lags behind many other countries, such as Brazil, Uruguay, Chile, Japan, South Korea, Israel, and the United Kingdom in the commercial use of UAS, particularly in agriculture.[9] This is noteworthy

---

[8] Rotorcraft with four, six, and eight rotors are commonly referred to as quadcopters, hexcopters, and octocopters, respectively.

[9] Y. Kim, "Grow More and Make It Affordable," presentation to the committee on November 13, 2013.

because agriculture is perhaps the largest potential market for UAS and is one major global economic activity where they can markedly lower costs and increase productivity. Led by Japan and Brazil, unmanned aircraft are used to precision-spray pesticides and fertilizers; more than 90 percent of crop dusting in Japan is done by unmanned aircraft such as the RMAX unmanned helicopter, which is remotely piloted without the aid of any autonomous systems. Japan's Administration of Agriculture oversees UAS operations there and requires annual and quarterly recurrent training of RMAX operators. Small unmanned aircraft equipped with cameras and other sensors can play another important role in agriculture by surveying crops to monitor growth, disease, and the application of pesticides, fertilizers, and water.[10] As a result, resources are applied where they are most needed, leading to production that is more economical and easier on the environment. Key technologies such as geofencing[11] enable aircraft to stay within the confines of a specified region, including altitude limits. Many of these algorithms are being developed in open source and are readily available on the Internet.[12]

The ability of UAS to operate in harsh, remote environments with no risk to human life facilitates environmental research; a remotely piloted Global Hawk unmanned aircraft was recently used in Canada for ground mapping the Arctic and studying weather phenomena.[13] Unmanned aircraft have also been used to monitor Japan's Fukushima nuclear power plant accident in places too dangerous for humans.

The entertainment industry is also capitalizing on the use of UAS; when equipped with high-quality cameras, UAS are a viable alternative to expensive and sometimes dangerous filming from crewed helicopters. In March 2013, Paramount Pictures used a swarm of unmanned aircraft to create a Star Trek emblem over the Tower Bridge in London to advertise an upcoming movie. Attention has recently focused on the use of unmanned aircraft for high-end real estate marketing by capturing aerial views of estates both inside and out. However, the FAA prohibits the commercial use of unmanned aircraft in the United States without proper certificates of waiver or authorization,[14,15] and it has shut down some innovative uses of UAS, such as on-site delivery of beer to ice fisherman by a brewery in Wisconsin.[16] In addition to these regulatory impediments, low-altitude operations in cluttered airspace present a very difficult operational challenge as well as a correspondingly large commercial opportunity for the use of small UAS.[17]

BENEFITS AND USES OF INCREASED AUTONOMY IN NONAVIATION APPLICATIONS

Aviation is by no means the only domain to exploit increased autonomous capability. The use of increased autonomy in several other application domains has been significant and will continue to influence the development of IA systems with relevance to civil aviation. Nonaviation applications of IA systems have been largely driven by a few key factors:

- Missions, functions, or tasks that cannot be performed by a human (e.g., real-time control of spacecraft during entry, descent, and landing on Mars).

---

[10] Associated Press, 2013, Agriculture the most promising market for drones, December 14, http://www.foxnews.com/us/2013/12/14/agriculture-most-promising-market-for-drones.

[11] A geofence uses GPS to check that an unmanned aircraft is within its designated operating area. If the aircraft approaches or exits this area, a return-to-base instruction is automatically executed to bring the aircraft back inside its designated operating area, which is defined by minimum and maximum altitude as well as by lateral (latitude and longitude) constraints.

[12] See, for example, Ardupilot (http://ardupilot.com), Arduplane (http://plane.ardupilot.com), and Openpilot (http://www.openpilot.org/).

[13] Northrop Grumman, NASA fly Global Hawk in Canadian airspace for first time to study Canadian Arctic, *PRNewswire.com*, December 19, 2013, http://www.prnewswire.com/news-releases/northrop-grumman-nasa-fly-global-hawk-in-canadian-airspace-for-first-time-to-study-canadian-artic-236562771.html.

[14] *New York Times,* Still unconvinced, home buyer? Check out the view from the drone, December 23, 2013, N.Y. Region, http://www.nytimes.com/2013/12/24/nyregion/still-unconvinced-home-buyer-check-out-the-view-from-the-drone.html?_r=0.

[15] FAA, 1981, Advisory Circular 91-57, Model Aircraft Operating Standards, June 9.

[16] Liz Fields, 2014, FAA slaps down drone beer delivery service to ice fishermen, *ABC News*, January 31, http://abcnews.go.com/US/faa-slaps-drone-beer-delivery-service-ice-fishermen/story?id=22314625.

[17] NASA, 2014, Enabling Civilian Low-Altitude Airspace and Unmanned Aerial System (UAS) Operations, Unmanned Aerial System Traffic Management (UTM) Workshop, NASA Ames, February 12-13.

- Nature of the mission or task (e.g., monitoring tasks that are so dull that it is difficult for humans to maintain a high state of alertness).
- Impacts and/or risks associated with relying on humans instead of IA systems.
- Economic and safety benefits.

Today's fielded autonomous systems are largely made up of subsystems capable of performing individual functions or tasks autonomously and are often not integrated into accomplishing more complex tasks. As a result, overall mission capabilities are often realized only with considerable human oversight. As in aviation, future systems are likely to be far more functionally integrated, requiring only high-level supervision on the part of human operators. Over time, these systems will likely focus on accomplishing more complex tasks in more uncertain environments and be capable of effective contingency management for a wide range of anomalous conditions. As a result, they will require less and less continuous human cognizance and control.

Ground Applications

The largest arena for the use of IA systems has been ground-based applications, covering a wide variety of stationary and mobile systems. Industrial manufacturing robots, consisting largely of stationary machines, were among the first to employ high levels of automation, having entered the marketplace in numbers 30 to 40 years ago. They have progressed from relatively simple, single-function machines (e.g., programmable milling machines) that required a well-defined operating environment to highly sophisticated multifunction machines capable of complex assembly and highly precise fabrication tasks under conditions of considerable uncertainty in geometry or environment. However, as sophisticated as these robotic manufacturing machines may be, they do not possess—nor do they need to possess—the sophisticated analytical capabilities that will be characteristic of the advanced IA systems for civil aviation.

Autonomy in the medical arena has been mostly evident in automated patient monitoring systems of various types. Although touted widely, medical robots today are mostly teleoperated systems guided by trained, skilled medical professionals who can add a high level of precision in certain surgical procedures.

Mobile systems (mobile robots, automobiles) that employ IA capabilities afford the greatest opportunity for synergy with the aviation world. Terrestrial mobile robots are used for diverse missions ranging from mine clearing and ordnance disposal to distribution of food carts in hospitals. The more sophisticated systems can perform complex physical tasks (obstacle avoidance and/or traversal, navigation in unknown environments, tactile object grasping, and autonomous inspection), while the simpler are typically focused on a single task and use relatively primitive algorithms. Most of these systems today involve a moderate to high level of human user monitoring or interaction. Robots for ordnance clearing and disposal employ sophisticated control and navigation systems traversing geographically challenging (uneven and uncertain) terrain, but they are largely teleoperated to accomplish their missions.

Automobiles have experienced the most remarkable growth in increased autonomy over the last decade. Since GM introduced the first electronic control unit in 1977, automobile control and monitoring systems have been transformed from almost purely mechanical units to the complex networks of microprocessors they are today. A typical luxury car now requires 70 to 100 electronic control units and over 100 million lines of computer code. This contrasts with 6.5 million lines of code for the avionics and onboard support systems on the Boeing 787 airplane.[18] Software and electronics may account for 40 percent of a new car's cost and 50 percent of warranty claims. This computer infrastructure that underlies modern vehicles enables a rapidly growing set of functions that includes IA cars.

These computers are already changing what it means to drive. They enable cars to take over many important driving operations, with features such as automatic parking and autonomous braking. Add to that lane-keeping assist and speed regulation systems, and the automation might exert more control of the vehicle than the driver. Already Volvo's City Safety automatic braking system has reduced injury-related crashes by 18 to 33 percent

[18] R. Charette, 2009, This car runs on code, *IEEE Spectrum,* February 1, http://www.spectrum.ieee.org/feb09/7649.

according to one estimate,[19] and the 2014 Mercedes S-class can drive at highway speed without input from drivers for short periods. The computer architectures in most production automobiles today feature federated systems, each of which specializes in the performance of individual automated functions, each one targeting specific attributes, such as fuel economy (e.g., fuel regulation) or safety (e.g., antilock brakes). In the last few years, mission-level autonomy has emerged to offer even more advanced capabilities, one of which is adaptive cruise control, which maintains vehicle-to-vehicle spacing.

The Google car initiative is further advancing and integrating these capabilities to provide a driverless, autonomous vehicle. Nissan has committed to the production of commercially available "autonomous" vehicles by the year 2020, but these vehicles will still require drivers to take action in some circumstances, and the full set of driver roles and responsibilities remains to be determined.[20] In any case, in the future a driver's primary task might not be driving but rather monitoring vehicle automation and interacting with information and entertainment systems. This shift introduces new vulnerabilities that can compromise driving safety.

As the technologies continue to mature with field testing and exposure to the driving public, they can be expected to gain increasing acceptance in a competitive marketplace that values economics, safety, novelty, and even prestige. Automobiles are not the only benefactors of IA capabilities: Trucks that operate without a driver could have obvious commercial value. But for many people, the attraction of autonomous systems lies in the chance for drivers to become passengers who can attend to work or enjoy new onboard entertainment and social networking systems. The safety benefits associated with some collision warning and collision mitigation systems, such as autonomous braking and stability control, also inspire the hope that vehicle-related fatalities can be greatly reduced.[21] As with civil aviation, the use of IA systems in ground vehicles is supported by rapidly improving sensor and computer technology.

The promise of IA cars faces several important challenges. These include complex hardware and software failures that traditional verification and validation (V&V) processes might not address, cybersecurity vulnerabilities that threaten vehicle control, electromagnetic interference that could corrupt signals, and human–machine integration issues.[22] Of these, the latter represent a particular challenge because one of the main appeals of IA vehicles is that they will relieve the driver of the need to continuously attend to the road and will enable him or her to engage in other activities. Although vehicle automation systems may accommodate the least demanding driving situations and encourage drivers to disengage from the driving task, they will also demand that a driver be able to reengage in the control loop, sometimes more quickly than is humanly possible. Drivers are most likely to fully rely on automation during long-duration highway driving, when a driver will need to very rapidly reenter the control loop if something suddenly goes wrong. This poses a safety challenge, a legal challenge regarding presumed responsibility, and a social challenge with respect to acceptance of the new technology.

Although ground vehicle applications of autonomy for the most part differ substantially from civil aviation applications, there are also several parallels and opportunities to build joint expertise on responding to the challenges of increased autonomy:

- The drivers of cars and the pilots of general aviation aircraft would both benefit from relatively inexpensive IA systems that could to some degree serve as copilots, alerting the driver or pilot of hazards that have been overlooked and taking corrective action in extremis.
- V&V of automobiles with IA systems face challenges similar to those faced by civil aviation. The National Highway Traffic Safety Administration (NHTSA) does not regulate or certify vehicle designs in the way that FAA regulates and certifies aircraft designs—and FAA certification standards tend to be much more

[19] Volvo, City Safety loss experience—An update, *Highway Loss Data Institute Bulletin* 29(23): 1.

[20] Nissan boss wants autonomous cars by 2020, *Autocar*, March 7, 2014, http://www.autocar.co.uk/car-news/geneva-motor-show/nissan-boss-wants-autonomous-cars-2020.

[21] Volvo, City Safety loss experience—An update, *Highway Loss Data Institute Bulletin* 29(23): 1.

[22] National Research Council, 2012, *The Safety Challenge and Promise of Automotive Electronics: Insights from Unintended Acceleration*, Transportation Research Board Special Report 308, Washington, D.C.: The National Academies Press.

rigorous than the safety standards that must be met before IA systems can be introduced into cars.[23] Even so, NHTSA and the FAA may benefit from sharing information regarding their efforts to assure that IA systems are safe and reliable. In particular, the FAA may benefit from operational data collected on the performance of advanced IA systems deployed in cars.

- Because both aircraft and ground vehicles operate in challenging physical, electromagnetic, and social environments, they may face similar vulnerabilities. These vulnerabilities range from unusual electrical failures (e.g., tin whisker shorts) to cyberphysical security.[24]
- Issues related to coordination of control in ground vehicles may be relevant to aircraft. The safety and performance of automobiles depend on the surrounding vehicles. The network effects of vehicle automation that propagate across a traffic stream could dominate vehicle response in some situations. This may become an even stronger influence with the connected vehicle concept, which provides data links among different vehicles and between vehicles and the infrastructure. Civil aviation may face a similar situation as aircraft communicate and interact more directly with one another.[25]

Marine Applications

Unmanned surface maritime vehicles are relatively new to the robotic family. Most contemporary systems are highly operator intensive (i.e., teleoperated), but they do exploit IA technologies such as augmented stabilization and control in high seas and during waypoint navigation. Advanced unmanned underwater vehicles are acquiring higher levels of autonomous functionality to enable extended periods of unattended operation. Employing autonomous navigation and other autonomous robotic functions, the relatively slow, deliberate movements and transit speeds of unmanned underwater vehicles help ensure that they have ample time to process information for autonomous decision making. Unmanned surface vehicles, which may operate at much higher speeds than underwater vehicles, also need to detect and avoid other watercraft. Marine vehicles operate in an environment with no inherent structure, such as roads.

Space Applications

Space missions traditionally rely heavily on humans for mission-level decisions, even in the case of robotic missions with sophisticated satellites or planetary exploration rovers. To reduce manpower requirements and account for the time delays in communications, the International Space Station (ISS) incorporates advanced smart sensors for failure recognition, diagnostics, and prognostics; model-based reasoning for scheduling maintenance; and automation of low-level routine tasks (e.g., directing imaging sensors). However, many advanced in-space functions such as inspection, maintenance, and assembly on board the ISS still require teleoperation or supervision by a human operator. IA technologies for deep space missions are required for tasks such as autonomous precision landing, and they greatly facilitate the operation of planetary rovers, enabling them to localize and avoid hazards and to execute functions that cannot be efficiently controlled from the ground because of communication delays. The Mars' Curiosity rover operates with scripted plans provided by ground personnel. The rover executes the plan using sensors to detect and control the vehicle to navigate terrain and avoid obstacles. Advanced technologies in space systems include real-time and life-cycle vehicle health information to enable advanced decision-making algorithms to inform logistics management and maintenance planning.

[23] NHTSA provides guidance to manufacturers through the Federal Motor Vehicle Safety Standards, which set minimum performance requirements but do not specify how manufacturers should meet those requirements. NHTSA assesses compliance by inspecting and testing sample vehicles. It also monitors for safety defects by examining consumer reports of anomalous vehicle behavior that might pose a safety risk.

[24] NRC, 2012, *The Safety Promise and Challenge of Automotive Electronics: Insights from Unintended Acceleration,* Transportation Research Board, Washington, D.C.: The National Academies Press.

[25] NHTSA is taking steps to enable vehicle-to-vehicle communication technology for light vehicles. It has recognized the potential of this technology to improve safety by allowing vehicles to communicate with each other and ultimately avoid many crashes altogether by continuously exchanging basic safety data, such as speed and position.

Space applications of autonomy may contribute new algorithms for a variety of functions that would benefit the civil aviation community. These include route planning, autonomous navigation, obstacle avoidance, autonomous landing site selection, and automation of lower-level maintenance and operational functions. Another contribution will come in the form of the crewing concepts needed to manage a constellation of vehicles versus the single-platform model now widely used. Space systems already employ a different crewing paradigm than typical crewed systems. Perhaps the most significant application of autonomy to civil aviation will come from the technologies used to deal with the long time delays between the ground and the spacecraft. The space domain has learned to use robust software algorithms to successfully operate spacecraft without continuous human inputs.

3

Barriers to Implementation

The committee has identified many substantial barriers to the increased use of autonomy in civil aviation systems and aircraft. These barriers cover a wide range of issues related to understanding and developing increasingly autonomous (IA) systems and incorporating them into the National Airspace System (NAS). Some of these issues are related to technology, some to certification and regulation, and some to legal and social concerns:

- Technology Barriers[1]
  —Communications and data acquisition
  —Cyberphysical security
  —Decision making by adaptive/nondeterministic systems[2]
  —Diversity of aircraft
  —Human–machine integration
  —Sensing, perception, and cognition
  —System complexity and resilience
  —Verification and validation (V&V)
- Regulation and Certification Barriers
  —Airspace access for unmanned aircraft
  —Certification process
  —Equivalent level of safety
  —Trust in adaptive/nondeterministic IA systems
- Additional Barriers
  —Legal issues
  —Social issues

Each of these barriers overlaps with one or more of the others, and efforts to overcome these barriers will not proceed in isolation one from the other. For example, the technology barriers tend to overlap with the certification barriers because advanced civil aviation technologies cannot be deployed operationally unless and until

[1] The committee did not prioritize the barriers; they are listed alphabetically within each group.
[2] Adaptive and nondeterministic systems are defined in the section "Decision making by adaptive/nondeterministic systems."

they can successfully complete the certification process. Likewise, advanced technologies cannot be expected to complete the certification process unless there are robust V&V processes to support technology development and certification.

The committee did not individually prioritize these barriers (for example, by urgency or degree of difficulty). However, the committee believes that there is one critical, crosscutting challenge that must be overcome to unleash the full potential of advanced IA systems in civil aviation. This challenge may be described in terms of a question: "How can we assure that advanced IA systems—especially those systems that rely on adaptive/nondeterministic software—will enhance rather than diminish the safety and reliability of the NAS?" There are four particularly challenging barriers that stand in the way of meeting this critical challenge:

- Certification process,
- Decision making by adaptive/nondeterministic systems,
- Trust in adaptive/nondeterministic IA systems, and
- Verification and validation.

TECHNOLOGY BARRIERS

Communications and Data Acquisition

Barrier Statement: Civil aviation wireless communications are fundamentally limited in bandwidth, and the operation of unmanned aircraft in the NAS could substantially increase the demand for bandwidth.

Wireless communications and data acquisition are foundational to the NAS. Very high frequency (VHF) radio and satellite communications are used for voice communication and data transmission among pilots, air traffic controllers, and airline operational controllers. Other segments of the frequency spectrum are used for land- and satellite-based navigation. Additional frequencies are dedicated to systems such as aircraft radar transponders that facilitate informed decision making and control within the NAS.[3] The Federal Communications Commission has allocated a fixed amount of bandwidth for civil aviation purposes. This amount is unlikely to increase. In fact, nonaviation commercial and consumer demand for bandwidth is increasing to the point where the use of some frequencies for aviation may be threatened.[4] Wireless devices (cell phones, tablets, and the like) are proliferating at an astounding rate, and the demand for bandwidth to support such devices is growing just as fast.

As the number of IA systems in operation increases, it will become more of a challenge to share the limited radio spectrum with existing legacy systems. The Federal Aviation Administration's (FAA's) roadmap for integrating unmanned air systems (UAS) in civil airspace has concluded that

> . . . harmonized radio spectrum is needed for UAS control and communications links to help ensure their protection from unintentional radio frequency interference, to help ensure adequate spectral bandwidth is available for meeting the projected command and control link capacity demands, and to facilitate operation of UAS across international borders. While spectrum is also needed for beyond-line-of-sight command and control links, the initial focus was on radio line-of-sight for civil UAS because demand for line-of-sight links is expected to be greater.[5]

The proliferation of remotely piloted UAS, as they are currently operated, will increase the demand for communications and data acquisition. These systems use communications/data links to connect remotely located pilots, aircraft, and air traffic controllers. Managing the transmission of video and airborne weather radar is key to maximizing available bandwidth. The demand for both of these high-bandwidth systems peaks during the departure

[3] FAA, "Spectrum Resources for Air Traffic Control Systems," Office of Spectrum Management and Policy, January 2002, http://www.faa.gov/about/office_org/headquarters_offices/ato/service_units/techops/spec_management/library/general.cfm.

[4] W.R. Voss, 2012, Gathering storm, *AeroSafety World*, February.

[5] FAA, 2013, *Integration of Civil Unmanned Aircraft Systems (UAS) in the National Airspace System (NAS) Roadmap First Edition—2012*, November 7, http://www.faa.gov/about/initiatives/uas/media/uas_roadmap_2013.pdf, p. 56.

and arrival phase of flight. Unmanned aircraft in a fully autonomous mode would greatly reduce the demand for bandwidth associated with aircraft operations. Even so, some UAS missions will involve real-time streaming video and/or sensor data, which could test the limits of available communications bandwidth.

Cyberphysical Security

Barrier Statement: The use of increasingly interconnected networks and increasingly complex software embedded throughout IA air- and ground-based system elements, as well as the increasing sophistication of potential cyberphysical attacks, threaten the safety and reliability of IA systems.

The NAS is a complex cyberphysical system composed of complex cyberphysical systems on aircraft, in air traffic monitoring (ATM) systems, in maintenance systems, and in other support systems. These cyberphysical systems have integrated computational and physical elements, including distributed microprocessors with embedded software, sensors, control systems, navigation, internal data networks, and external communications. The complexity of these ever more interconnected systems can create multiple attack vectors for cyberthreats.

There are many positive motives and drivers for the growing complexity of the NAS. Embedded processors throughout the aircraft are used to improve operational performance, add important safety features, and perform automated vehicle health and maintenance functions. Higher levels of automation/autonomy drive the need for additional software, processors, and sensors throughout the system. Additional communications, wireless connectivity, and in-flight entertainment options are also adding to the complexity of these systems. The NextGen system[6] offers the promise of greater capacity, operational safety, and efficiency, but it also increases networking and system-of-system complexity. The need for additional research and development to address the challenges for software V&V was noted in the *NextGen UAS Research, Development and Demonstration Roadmap,*[7] but cyberphysical threats were not addressed.

In conventional desktop and network cyberenvironments, there is a growing asymmetry between the typical code size and attack surface of application software versus the relatively small size of typical software viruses. In some cases, even antivirus software itself can create new vulnerabilities. On a recent software vulnerability watch list, about one-third of the reported software vulnerabilities were in the security software itself.[8]

Today's modern aircraft have millions of lines of computer code embedded in distributed processors throughout the aircraft. Cyber vulnerabilities can arise from the reliance of aircraft systems on sensors, control systems, navigation, communications, and other physical elements, and from their interaction with other air and ground systems. There have been numerous documented examples of intentional and unintentional GPS jamming and spoofing. As the NAS transitions to the NextGen system and expands the use of Automatic Dependent Surveillance–Broadcast (ADS-B) systems,[9] there will be increased reliance on onboard sensing such as GPS for efficient and safe operation. Communications is also a key vulnerability for cyberphysical systems. Radio hackers have used air traffic control frequencies to give pilots false commands.[10] Communications systems are of increased importance for UAS because they provide the link to the human operators. As machine-to-machine communications share more information on aircraft navigation, health, and other data, there will be additional opportunities to exploit these network connections. The embedded processing elements themselves can be a point of vulnerability, with the threat of tampering with the chips to add hidden functionalities and unsecure backdoor access.[11]

---

[6] The Next Generation Air Transportation System (NextGen) is the a new National Airspace System due for implementation across the United States over the next decade.

[7] NextGen UAS Research, Development and Demonstration Roadmap, Version 1.0, March 15, 2012, http://www.jpdo.gov/library/20120315_UAS%20RDandD%20Roadmap.pdf.

[8] DARPA High-Assurance Cyber Military Systems (HACMS) Proposer's Day Brief.

[9] ADS-B is a key component of the NextGen system. The ADS-B system on an aircraft continuously broadcasts that aircraft's position and other flight information, all of which can be displayed on other aircraft and on ground systems.

[10] *Aerospace America,* Getting ahead of the threat: Aviation and cyber security, July-August 2013, pp. 22-25.

[11] *The Guardian*, Cyber-attack concerns raised over Boeing 787 chip's back door, May 29, 2012.

Decision Making by Adaptive/Nondeterministic Systems

Barrier Statement: The lack of generally accepted design, implementation, and test practices for adaptive/nondeterministic systems will impede the deployment of some advanced IA systems and aircraft in the NAS.

Adaptive systems have the ability to modify their behavior in response to their external environment. For aircraft systems, this could include commands from the pilot and inputs from aircraft systems, including sensors that report conditions outside the aircraft. Some of these inputs, such as airspeed, will be stochastic because of sensor noise as well as the complex relationship between atmospheric conditions and sensor readings that are not fully captured in calibration equations. Adaptive systems learn from their experience, either operational or simulated, so that the response of the system to a given set of inputs varies over time. This variation means that traditional verification, validation, and certification (VV&C) methods are incompatible with adaptive systems, since those methods are predicated on achieving a predictable output, albeit with some level of measurement error, for a given input. New approaches to VV&C are required to demonstrate that adaptive systems will consistently make appropriate decisions even though individual test results may not be repeatable. The expectation that adaptive systems will improve over time is expected to be of such value that research into these systems is worthwhile despite the need for new VV&C methodologies.

Systems that are nondeterministic may or may not be adaptive. They may be subject to the stochastic influences imposed by their complex internal operational architectures or their external environment, meaning that they will not always respond in precisely the same way even when presented with identical inputs or stimuli. The resulting uncertainty means that traditional VV&C methods are incompatible with nondeterministic systems. VV&C methods for complex, nondeterministic systems may be even more difficult to develop than VV&C methods for adaptive systems. However, the software that is at the heart of nondeterministic systems is expected to enable improved performance because of its ability to manage and interact with complex "world models" (large and potentially distributed data sets) and execute sophisticated algorithms to perceive, decide, and act in real time. As with adaptive systems, the improved performance of nondeterministic systems is expected to be of such value that research into these systems is worthwhile despite the need for new VV&C methodologies.

Systems that are adaptive and nondeterministic demonstrate the performance enhancements—and the VV&C challenges—of both characteristics. Many advanced IA systems are expected to be adaptive and/or nondeterministic, and they should be evaluated on the basis of expected or intended responses to a representative set of input conditions to ensure that they will consistently produce an acceptable outcome under a wide variety of environmental conditions and operational situations. This requires an understanding of how these systems sense and perceive internal and external data and the rationale by which the system arrives at its output decision. Existing adaptive/nondeterministic algorithms have not been widely applied to safety-critical civil aviation applications in part because of the lack of a mature process for designing, implementing, and testing such algorithms.

Diversity of Aircraft

Barrier Statement: It will be difficult to engineer some IA systems so that they are backward-compatible with legacy airframes, ATM systems, and other elements of the NAS.

Many civil aircraft have life cycles extending to decades. In general, new aviation technologies are required to be backward compatible with legacy aircraft and systems. Integration of advanced systems, such as IA systems, requires consideration of the size, missions, and capabilities of legacy systems now and well into the future.

Legacy systems include aircraft that do not have transponder or voice radio capabilities and sailplanes that operate without engines. More typical legacy aircraft have radios, transponders, and engines, but some of their system capabilities may be out-of-date by several decades compared to those of new systems coming off the production line. Some advanced aviation systems—and this is likely to include many IA systems—work most efficiently when interconnected with other similarly equipped aircraft. However, IA systems will also need to be designed and engineered so that they can operate safely even in the absence of those connections and, preferably, without imposing new tasks or requirements on human pilots in legacy aircraft.[12]

Human–Machine Integration

Barrier Statement: Incorporating IA systems and aircraft in the NAS would require humans and machines to work together in new and different ways that have not yet been identified.

There will always be some level of human involvement in the operation of IA systems, if for no other reason than to initiate and/or terminate otherwise autonomously performed tasks or missions. Thus, there will always be some kind of interface that allows humans to communicate intent to the IA system and that allows the system to communicate appropriate state and status information to the human. Although many of the necessary characteristics and requirements of systems that will enable humans to interact effectively with IA systems are known from previous research on human–automation interaction, much remains to be learned as we transition to systems with ever-increasing levels of autonomy.

Certainly one fundamental aspect to consider is the distribution of roles, responsibilities, and workload among human operators in IA systems. Implementation of IA systems will be adversely impacted if the workload is distributed poorly between humans and machines, either by depriving humans of necessary information or by overloading them with too much information in too short a time frame. To be effective, autonomous components should be designed to be competent players in an overall system that teams humans with machines. Critical analysis to define the appropriate functional allocation of the roles between the systems and the humans will be essential to avoid design pitfalls that would, in effect, require the human to be the ultimate fail-safe mechanism if the autonomous elements fail.

Effective integration of IA systems into the NAS will require consideration of the impacts of these operations on all stakeholders, including legacy aircraft and systems. For example, traffic management and collision avoidance in mixed operations of legacy aircraft and aircraft with IA systems will require transparency so that all responsible agents, human and machine, will understand the intentions and plans of the others. This presents the considerable challenge of designing human–machine interfaces that will enable the coordination of such operations within confined airspace.

Experience shows that human operators have a tendency to become reliant on the continued functionality and proper performance of automated/autonomous systems. This can be particularly problematic when the performance of these systems degrades, sometimes subtly, sometimes obviously, and sometimes suddenly. Overreliance on automation is frequently suspected as a factor in or indeed the cause of aviation incidents and accidents. If the manual skills necessary to perform an operation or maneuver have not been sufficiently practiced and maintained, human operators may lose proficiency and be ill-prepared to take over in the event of system failures. Overreliance and loss of proficiency can each make operators reluctant to assume manual control even in the face of clearly faulty automation.

Conventional automated flight systems sometimes experience gradual failures without obvious indications, which can make it difficult for operators to detect faults and prepare themselves to take over manual control at an appropriate time. Proper human–machine integration for IA systems would promote graceful degradation of system performance, whereby operators are informed of impending or potential failure modes that may require human intervention.

Formal training programs contribute to the NAS's high level of safety. These programs have been developed through many years of experience and have evolved in response to lessons learned from accidents and incidents, new applications of advanced automation/autonomy, and new training technology such as high-fidelity simulators. Appropriate training requirements and programs for IA systems will need to be established before the systems can be deployed.

---

[12] Boeing Commercial Airplanes, "Boeing Autonomy Research Recommendations," presentation to the committee, November 13, 2013.

Sensing, Perception, and Cognition

Barrier Statement: The ability of IA systems to operate independently of human operators is fundamentally limited by the capabilities of machine sensory, perceptual, and cognitive systems.

Machine sensing uses sensors to detect and extract information about the internal states of the system and the external states of the operating environment. Machine perception is the transformation of this raw sensor data into high-level abstractions that can be interpreted by machines and/or understood by humans and used to support decision making. Machine cognition is the utilization and application of this processed information to recognize patterns and make decisions relevant to the missions or functions being performed. All three processes are central to the ability of machines to operate independently of human operators and are often described by use of the OODA loop (Observe, Orient, Decide, and Act), discussed in Chapter 1.

Machine perception is currently very difficult for some combinations of sensors and tasks. For example, while it is relatively easy to detect the presence of an object in a specific location, it is much more difficult to identify or characterize that object (for example, as a building or a vehicle), and it is even more difficult to "contextualize" it (landing pad, conflicting traffic). Projecting the future state or actions of a detected object is still more daunting. Advances in machine perception capabilities are therefore crucial to the implementation of adaptive/nondeterministic IA systems.

In some circumstances it might be beneficial to share onboard sensor data among multiple aircraft. In these situations, the sensors on board each aircraft become part of a distributed sensor network that can combine data to provide better information than any single aircraft can achieve on its own. However, networking sensors can also introduce communication lags and interference, which may decrease precision. As a result, distributed sensor networks may have performance profiles that differ dramatically from more traditional onboard sensors.[13]

Machine cognition is a major challenge for the development of IA systems. High-level cognitive ability of machines is necessary to enable them to operate independently of humans for complex, dynamic missions. Achieving this will require consideration of the expectations of the humans who are managing these systems, especially if the system design includes provisions for some level of human intervention in the event of degraded performance or partial system failures.

System Complexity and Resilience

Barrier Statement: IA capabilities create a more complex aviation system, with new interdependencies and new relationships among various operational elements. This will likely reduce the resilience of the civil aviation system, because disturbances in one portion of the system could, in certain circumstances, cause the performance of the entire system to degrade precipitously.

The report produced by the Defense Science Board (DSB) Task Force on Autonomy notes that as capabilities of IA platforms continue to grow, design and testing activities should be focused on the larger multirole, multi-echelon system in which the vehicle operates.[14] For example, the report states that "a key challenge presented by the complexity of software is that the design space and trade-offs for incorporating autonomy into a mission are not well understood and can result in unintended operational consequences." The Networking and Information Technology Research and Development (NITRD) report on Grand Challenges[15] notes the opportunity to "understand how people, (software) agents, robots, and sensors (PARS) contribute to a collaboration" and the difficulties of

---

[13] R. Olfati-Saber, J.A. Fax, and R.M. Murray, 2007, Consensus and cooperation in networked multi-agent systems, *Proceedings of the IEEE* 95(1): 215-233.

[14] Defense Science Board, 2012, *Task Force Report on the Role of Autonomy in DoD Systems,* Office of the Secretary of Defense, July, http://www.acq.osd.mil/dsb/reports/AutonomyReport.pdf .

[15] Networking and Information Technology Research and Development Program (NITRD), 2006, *Grand Challenges: Science, Engineering, and Societal Advances Requiring Networked Information Technology Research and Development,* November, Washington, D.C., http://www.nitrd.gov/Publications/PublicationDetail.aspx?pubid=43, p. 31.

understanding "the structural complexity of PARS collaborations (for example, teams, networks, or hierarchies into which the PARS components can self-organize)." Echoing the DSB findings, the NITRD report calls for research and development on architectures for adaptive layered networks. A 2012 NITRD workshop report directly addresses the complexity issues and notes that the technical means to design and test complex engineered networked systems do not yet exist.[16] That report points to the need to develop the ability to design networks of IA vehicles and humans that demonstrate increased flexibility, robustness, resilience, and extensibility even as the operating environment, technology, and applications change over time.

The DSB report cautions that "current designs of autonomous systems, and current design methods for increasing autonomy, can create *brittle* platforms, and have led to missed opportunities and new system failure modes when new capabilities are deployed. Brittle autonomous technologies result in unintended consequences and unnecessary performance trade-offs, and this brittleness, which is resident in many current designs, has severely retarded the potential benefits that could be obtained by using advances in autonomy." The report recommends the "development of measures and models of the dimensions of system resilience/brittleness that can be used early in systems development as well as later in T&E." The problem of brittle IA systems has been noted for years in human–machine systems research,[17,18] and overcoming brittleness is a critical goal that guides the development of the growing field of resilience engineering.[19]

The barrier of complexity arises because current designs tend to focus on onboard autonomous capabilities and downplay the need for supporting coordination in a distributed and layered system. Complex engineered networks are broader in perspective—that is, they comprise a multirole, multiechelon system that includes both human and various computational and robotic roles. They present a challenge in terms of their ability to synchronize the distributed activities to keep up with the changing pace and tempo of dynamic situations. The distributed system has to be highly adaptive—changing the relationships across roles and echelons to be able to match changing demands and situations—if it is to meet the robustness and resilience goals as well as the cost and productivity goals, especially as growth in platform autonomy creates new opportunities. Complex engineered networks pose challenges for distributed architectures, such as accommodating new and multiple scales, managing life-cycle extensibility and adaptability, managing multiple trade-offs adaptively, and handling the extensive dependencies created by reliance on software-intensive systems and vital digital infrastructure. Together these sources indicate that

- The growth of IA systems leads to the need for new models, measures, architectures, and testing and certification methods for complex engineered networks,
- Complex engineered networks present many unsolved technical challenges, and
- There are several promising trends in interdisciplinary research such as resilience engineering, resilient control systems, and cyberphysical systems.

Verification and Validation

Barrier Statement: Existing verification and validation (V&V) approaches and methods are insufficient for advanced IA systems.

V&V are critical steps on the path to operational acceptance. Verification refers to the processes for assuring that a given product, service, or system meets its specifications. Validation refers to the processes for assuring that the product will fulfill its intended purpose. Most often, the entity that sets the specifications performs the verifica-

[16] NITRD, 2012, *Workshop Report on Complex Engineered Networks,* September 20-21, Washington, D.C., http://scenic.princeton.edu/NITRD-Workshop/.

[17] P.J. Smith, C.E. McCoy, and C. Layton, 1997, Brittleness in the design of cooperative problem-solving systems: The effects on user performance. *IEEE Transactions on Systems, Man, and Cybernetics* 27(3): 360-371.

[18] S.J. Perry, R.L. Wear, and R.I. Cook, 2005, The role of automation in complex system failures. *Journal of Patient Safety* 1(1): 56-61.

[19] E. Hollnagel, D.D. Woods, and N. Leveson, eds., 2006, *Resilience Engineering: Concepts and Precepts,* Aldershot, U.K.: Ashgate Publishing.

tion, while the prospective user or some other organization that thoroughly understands user needs performs the validation.[20]

V&V methodologies and techniques are typically underpinned by inviolate scientific principles. For example, physics and or other related engineering sciences are most often used to determine whether an aeronautical system is capable of fulfilling its specifications and its intended purpose. Thus, V&V processes currently employed in aviation (and most other physical applications) are geared toward obtaining quantitatively predictable outcomes based on known inputs or stimuli. Specifically, to be viable, the selected architecture, instantiated with appropriate levels of sensing capability, must prove to be robust over a range of complex and uncertain operating environments. One logical alternative to the first-principles-based approach is to conduct a statistical evaluation (with associated confidence levels) of the behavior of the product, service, or system. However, even a statistical evaluation of system behavior may not be adequate for V&V of autonomous systems.

If the capability in question is accomplished through software, the creators of that software are generally expected to prove that the software can deliver the intended capabilities at specified performance levels and in the relevant environment. In civil aviation applications, software V&V are a precursor to achieving certification, which in turn is required for flight operations. To this end, the FAA requires that software developers follow the guidance contained in DO178B, *Software Considerations in Airborne Systems and Equipment Certification*.[21,22] Approaches to software V&V such as those described in DO178B are laborious because they require extensive examination and testing of every logic path. This approach is based on the well-established premise that errors may occur at any level. However, it is increasingly apparent that, given the complexity associated with software that has been and is being developed to provide increased functionality, these approaches are not scalable to the exceedingly complex software associated with some advanced IA systems. An alternative approach is required, especially at the system level.

In twentieth century aviation, system-level "intelligence" resided almost exclusively with human operators. The segregated assessment of the aircraft and crew for flight readiness was acceptable, because the standards were for the most part independent. That is not the case with IA systems, where the intelligence is likely to be divided between the human and the system. Even if a V&V method were to prove useful for assessing intelligent software, the assessment of the total system, including the human operators and their interactions with the IA system, requires another approach. The barrier here is one of developing confidence in a system with a significant variation in the degree of human involvement, capability, and management. To this end, the validation process will likely require IA systems and human operators to respond quickly and correctly to a great number of questions and situations.

Another element of this barrier relates to the difficulty of establishing system-level performance requirements and functionality that are applicable to and derived from specified levels of safety, reliability, and operational performance. In the case of advanced IA systems, these attributes would be specifically related to the intent of the systems, the mission and operating environment, and the level of human interaction and uncertainty. A generalized approach to overcoming this part of the barrier does not currently exist.

Lastly, new approaches to V&V processes are likely to affect system design processes, especially for small unmanned aircraft, where the cost of V&V of IA systems will need to be modest in order for the aircraft to be affordable.

REGULATION AND CERTIFICATION

Airspace Access for Unmanned Aircraft

Barrier Statement: Unmanned aircraft may not operate in nonsegregated civil airspace unless the FAA issues a certificate of waiver or authorization (COA).

[20] Institute of Electrical and Electronic Engineers (IEEE), 2008, *1490 WG-IEEE Guide: Adoptions of the Project Management Institute (PMI) Standard: A Guide to the Project Management Body of Knowledge (PMBOK Guide) Working Group*, 4th ed., http://standards.ieee.org/develop/wg/software_and_systems_engineering_all.html.

[21] FAA, 1993, Advisory Circular 20-115B, RTCA, Inc., Document RTCA/DO-I 78B, January 11.

[22] FAA, 2011, Advisory Circular 20-171, Alternatives to RTCA/DO-178B for Software in Airborne Systems and Equipment, January 19.

Under current regulations there is no routine way to accommodate unmanned aircraft operations in non-segregated civil airspace. Such operations have been possible only with (1) a certificate of authorization issued by the FAA on a case-by-case basis to public organizations or (2) a special airworthiness certificate in the experimental category. These certificates include operational restrictions that limit the utility and mission effectiveness of the affected unmanned aircraft. The FAA's cumbersome regulatory mechanisms for authorizing UAS operations will not scale to the volume of operations expected in the near future, and they will continue to constrain potentially beneficial operations. However, the status quo is changing. Two small unmanned aircraft, the InSitu Scan Eagle and the AeroVironment Puma, were certified for commercial usage (in restricted airspace) in September 2013, and the FAA subsequently approved commercial UAS operations, although this authorization was limited to very remote airspace (in the Arctic). In addition, the FAA has developed a roadmap for integrating UAS into the NAS with milestones for the FAA, other government agencies, and industry.[23] The FAA is also developing enabling regulations for small unmanned aircraft that are expected to be released for public comment in November 2014. The FAA policy restricting the commercial use of small unmanned aircraft is being challenged in court. On March 6, 2014, a National Transportation Safety Board (NTSB) administrative law judge ruled that the FAA's de facto ban on commercial use of small unmanned aircraft (model aircraft), which is based on a 2007 policy statement, "cannot be considered as establishing a rule or enforceable regulation" and that "policy statements are not binding on the general public." The ruling is under appeal.

Most unmanned aircraft presently in operation are remotely piloted by a human operator on the ground who is in continuous operational control unless there is a failure or disruption of the command-and-control link. The human operator is assisted by automation that can maintain stabilized flight by controlling aircraft attitude and power. In more highly automated aircraft, the pilot simply defines the desired flight path by specifying three- and four-dimensional waypoints, and the path is then executed by automation on board the aircraft.

The FAA says that "ultimately, UAS must be integrated into the NAS without reducing existing capacity, decreasing safety, negatively impacting current users, or increasing the risk to airspace users or persons and property on the ground any more than the integration of comparable new and novel technologies."[24] The committee believes that these are appropriate and achievable goals. In particular, the committee believes that the impact of IA systems on the performance of crewed aircraft, unmanned aircraft, and ATM systems will make it possible to introduce UAS into the NAS without reducing the capacity for crewed aircraft operations.

Full integration of UAS into civil airspace is limited by a variety of technical, operational, policy, and economic issues. The most daunting airspace integration challenges are these:

- *Detect-and-avoid capabilities.* Because unmanned aircraft do not have a pilot on board, the ability to see and avoid other aircraft is dependent on the development of new technologies.
- *Vulnerabilities of the command and control link.* The command-and-control link between the pilot and the aircraft is via a wireless communications link, which is subject to all the vulnerabilities of such links.
- *ATM integration.* The unique operational characteristics and flight performance of unmanned aircraft require UAS-specific ATM procedures and technologies.[25]

Certification Process

Barrier Statement: Existing certification criteria, processes, and approaches do not take into account the special characteristics of advanced IA systems.

Certification of aircraft and other civil aviation systems is critical to maintaining safety, especially in the midst of growth and/or changes in the operating environment. To date, there are no commonly accepted methods,

---

[23] FAA, 2013, *Integration of Civil Unmanned Aircraft Systems (UAS) in the National Airspace System (NAS) Roadmap, First Edition—2012,* November 7, http://www.faa.gov/about/initiatives/uas/media/uas_roadmap_2013.pdf, p. 56.

[24] Ibid, p. 4.

[25] Andrew Lacher, Andrew Zeitlin, Chris Jella, Charlotte Laqui, Kelly Markin, 2010, *Analysis of Key Airspace Integration Challenges and Alternatives for Unmanned Aircraft Systems,* F046-L10-021, McLean, Va.: The MITRE Corporation, July.

techniques, or practices for the certification of highly autonomous systems. Further, once new methods, techniques, and practices have been developed, they must be authorized by changes to FAA regulations; to the FAA Advisory Circulars that will alert the public on how the agency interprets and will apply new regulations; and to the guidance that is provided to certification specialists and other interested parties via FAA Orders on how to apply the new regulations. Developing new regulations, policies, procedures, guidance material, and training requirements, however, is lengthy and resource intensive, especially when new technologies involve expertise that is not currently resident in the staff of the responsible FAA offices. However, this situation has been encountered and overcome in the past as other revolutionary new technologies, such as fly-by-wire flight controls and composite materials, were introduced into civil aviation.

Having a properly certificated aircraft does not entitle one to operate it in U.S. airspace, especially if it is done for commercial purposes. Operation of an aircraft may require certification of the company that will be operating the aircraft, and it certainly requires pilot certification, whether the pilot is on board or flying the aircraft remotely. The certification of UAS "operators" (as a class of individuals distinct from pilots) to remotely operate unmanned aircraft is not contemplated by present regulations and would likely need major changes to the FAA operating rules. Such changes, as with certification rule changes, would also require the development of Advisory Circulars and FAA Orders.[26]

One of the most important aspects of new guidance documents for IA systems will be their treatment of the human factors aspects of IA aircraft operations. As first articulated in the 1996 FAA report on flight deck human factors and most recently updated in the recently released Flight Deck Automation Working Group report,

> Pilots mitigate safety and operational risks on a frequent basis, and the aviation system is designed to rely on that mitigation. . . . Automated systems have been successfully used for many years, and have contributed significantly to improvements in safety, operational efficiency, and precise flight path management. However, pilot use of and interaction with automated systems were found to be vulnerable in the following areas:
>
> - Pilots sometimes rely too much on automated systems and may be reluctant to intervene,
> - Autoflight mode confusion errors continue to occur,
> - The use of information automation is increasing, including implementations that may result in errors and confusion, and
> - FMS [Flight Management System] programming and usage errors continue to occur.[27]

Certifying entities will need to acquire and maintain trust in the judgments that adaptive/nondeterministic components and systems exercise. The certification community does not yet know how to do this or which specific criteria should be applied. Appropriate reliability metrics and acceptable standards for software development have yet to be identified. Nor has the community developed quantitative levels of system reliability and how these requirements could be allocated between hardware, software, and pilots. Nor are there available specific guidelines for redundancy management, which will surely involve new approaches to establish failure modes and effects analyses as well as the setting of specifications for performance of systems, sensors, and software and hardware testing and robustness. In addition, some of the unresolved certification challenges are of particular interest to rotorcraft. Examples include single-pilot workload evaluation, man–machine issues in the near-earth environment, and adaptation of functional allocations to single-pilot operations.

[26] Advisory Circulars and informational documents are issued by the FAA to provide background, explanatory, and guidance material to the aviation community for complying with regulations. FAA Orders are issued by the FAA to provide guidance to its inspectors, controllers, and other agency employees in the conduct of their respective duties.

[27] FAA, 2013, *Operational Use of Flight Management Systems,* Report of the PARC/CAST Flight Deck Automation Working Group, http://www.faa.gov/about/office_org/headquarters_offices/ avs/offices/afs/afs400/parc/parc_reco/media/2013/130908_PARC_FltDAWG_Final_Report_Recommendations.pdf.

Equivalent Level of Safety

Barrier Statement: Many existing safety standards and requirements, which are focused on assuring the safety of aircraft passengers and crew on a particular aircraft, are not well suited to assure the safety of unmanned aircraft operations, where the primary concern is the safety of personnel in other aircraft and on the ground.

Equivalent safety implies that the use of a new system, such as one that possesses advanced IA systems, will demonstrate the same level of safety as traditional aircraft operations. The regulatory structure that underpins safety of flight is the result of an evolution that has its origins in the early crewed aircraft era and is crafted around the notion of a pilot in the cockpit. The first generation of modern UAS itself evolved from this model, being heavily dependent on interactions between a remote human pilot and onboard systems with extremely limited autonomous capabilities. IA systems can enable the operation of unmanned aircraft with reduced dependence on the remote crew and less need for highly reliabile, real-time communications.

The current air safety regulatory framework implicitly assumes that the "system intelligence" resides exclusively in the human component of the system, primarily the pilots and air traffic controllers. Current automation within civil aircraft is focused on the execution of specific physical control tasks, such as stabilization and orientation, autotakeoff, autolanding, and autonavigation. Although some of these tasks may be complex, they are all inherently based on the execution of the OODA functions by air and ground elements of the NAS in a way that produces deterministic, predictable, repeatable outcomes. Today's regulatory framework assumes that the pilot is solely responsible for how the onboard automation is used.

In contrast, advanced IA systems derived from rapidly evolving intelligent machine technology will be capable of mission-level decisions comparable to those made by the pilot or other aircrew. Decision logic will be cued and driven by the system's perception, itself the product of inputs from sensors and other available information sources, along with instantiated and learned behaviors. For many UAS and their missions, the IA contribution to decision making will be an imperative rather than an option because of the real-world issues inherent in remote operation. To assure trust in IA systems, the regulatory system would therefore need to evolve to recognize and evaluate system performance in terms of how humans and machines work together in accomplishing their mission.

Trust in Adaptive/Nondeterministic Systems

Barrier Statement: Verification, validation, and certification are necessary but not sufficient to engender stakeholder trust in advanced adaptive/nondeterministic IA systems.

Successful deployment of IA systems will require mechanisms for establishing and maintaining trust in the systems' ability to perceive relevant environmental circumstances and to make acceptable decisions regarding the course or courses of action. One important characteristic of IA systems is that they will be able to cope with a range of both known and unknown inputs without frequent or nearly continuous human cognition and control. Thus, it will be important for IA systems to incorporate pattern recognition, hypothesis testing, and other functionalities that are typically performed by humans. For example, IA systems will need to correctly perceive environmental and internal state variables and then apply machine reasoning and judgment to make appropriate decisions and take appropriate courses of action.

Trust is not an innate characteristic of a system; it is a perceived characteristic that is established through evidence and experience. To establish trust in IA systems, the basis on which humans trust these systems will need to change—for example, by developing trust in them in a fashion analogous to how people establish and maintain trust in other people or organizations. Much of this trust is based on a person's confidence that others will behave in certain ways (through action or, in some cases, through carefully considered inaction), even under challenging and unforeseen circumstances. That is, building trust involves the estimation (with associated confidence) of both the bounds of performance of the system and the bounds of the control actions that will be performed within and by the system. This is in contrast to exhaustive testing over given operating conditions, as is typically now the case in VV&C.

After every incident or accident involving IA systems, it will be necessary to undertake a forensic evaluation of the performance of these systems, including their ability to appropriately perceive environmental circumstances and make suitable judgments. Evaluations could include formal testing, simulation, and limited operational deployments. Continued monitoring will be necessary to maintain confidence in adaptive/nondeterministic IA systems as they adapt to changes in the environment and feedback from previous experience. Establishing mechanisms to limit possible actions by IA systems will be necessary to effectively control the consequences of poor performance. Increasing levels of authority can follow as confidence grows in the ability of IA systems to perform appropriately. This is very similar to how extended-range twin operations (ETOPS) have been managed: As confidence in the reliability of engines and other aircraft systems has grown, the FAA has granted air carriers the authority to operate twin-engine commercial transports along routes with flight segments that are increasingly distant from airports to which such aircraft could be diverted in the case of an engine malfunction en route.

ADDITIONAL BARRIERS

Legal Issues

Barrier Statement: Public policy, as reflected in law and regulation, could significantly impede the degree and speed of adoption of IA technology in the NAS.

The FAA Modernization and Reform Act of 2012 included a congressional mandate to integrate UAS into the NAS by September 2015.[28] On March 6, 2014, an NTSB administrative law judge ruled that the FAA's de facto ban on commercial use of small unmanned aircraft (i.e., model aircraft) is not binding on the general public. (It remains to be seen how this this ruling will fare during the appeals process.) In any case, the commercial use of UAS in the NAS raises a host of legal issues. A Congressional Research Service report[29] identifies several legal issues, including constitutional concerns. The First Amendment guarantee of freedom of the press will be an important consideration as journalists increasingly employ UAS to gather information. Fourth Amendment protections regarding unreasonable searches conducted using UAS have yet to be established. UAS also raise new questions related to the Fifth Amendment prohibition against property "being taken for public use, without just compensation." In this context, such "taking" involves government interference in the ability of people to use UAS over their own property.

Government actors and private citizens may be held accountable under laws relating to trespass, nuisance, stalking, harassment, privacy, and liability. How these legal concepts are applied to UAS will be determined by legislative and legal action at the state and federal level. An indeterminate period of uncertainty will persist as the legal system adapts to the introduction of UAS having various operating profiles into the NAS for various applications.

Liability may prove the most significant legal challenge to unmanned aircraft. As advances enable more fully autonomous operations, human interaction with UAS will become more remote. The legal system will be challenged to account for systems causing harm based on decisions made without direct human input.

Between 2010 and 2012 the U.S. Customs and Border Protection Agency loaned its Predator UAS to other law enforcement agencies on 700 occasions.[30] On January 14, 2014, Rodney Brossart became the first American convicted of a crime in which surveillance video from an unmanned aircraft was presented as evidence at trial. In 2013, bills or resolutions addressing unmanned aircraft were introduced in the legislative bodies of 43 states.[31] Topics included limitations on government and personal use of unmanned aircraft and funding for FAA test sites. In the absence of legislative guidance, courts will adapt existing laws to address concerns arising from the use of unmanned aircraft. It is inevitable that the law covering IA systems in the NAS will continue to evolve.

---

[28] FAA, Modernization and Reform Act of 2012, P.L. 112-95, 126 Stat. 11.

[29] Congressional Research Service, 2013, *Integration of Drones into Domestic Airspace: Selected Legal Issues,* April 4, http://www.fas.org/sgp/crs/natsec/R42940.pdf .

[30] *U.S. News and World Report,* 2014, North Dakota man sentenced to jail in controversial drone-arrest case, January 15, http://www.usnews.com/news/articles/2014/01/15/north-dakota-man-sentenced-to-jail-in-controversial-drone-arrest-case.

[31] National Conference of State Legislatures, "2013 unmanned Aerial Systems (UAS) Legislation," http://www.ncsl.org/research/civil-and-criminal-justice/unmanned-aerial-vehicles.aspx.

Even as advanced IA systems are deployed in the NAS, IA systems—like other aviation technologies—will continue to mature. As the operational record of advanced IA systems grows, it is likely to trigger additional reviews of policy, laws, and regulations that govern their use.

Social Issues

Barrier Statement: Social issues, particularly public concerns about privacy and safety, could significantly impede the degree and speed of adoption of IA technology in the NAS.

Civil aviation provides many benefits to the general public, such as the safe transport of millions of passengers annually, overnight delivery of packages, and emergency medical evacuation of accident victims. However, the opinions of the general public about technology are shaped by a wide variety of forces and factors. For example, the intense coverage of aviation accidents can create the impression that commercial air travel is less safe than objective measures indicate. Noise concerns, particularly around busy airports, can also color the public's impressions of aviation.

The general public has limited detailed technical knowledge about the NAS. Most people do not know how air traffic is routed or coordinated. Nor do they know the technical challenges that had to be overcome to assure the safety of civil aviation as new technologies such as fly-by-wire flight controls and composite materials came to play vital roles in the design and operation of commercial transports. With two noteworthy exceptions, the same will likely be true when it comes to the application of IA systems in civil aviation.

First, unmanned aircraft are visibly different from crewed aircraft, and the military uses of drones for surveillance and active warfare are frequently discussed in print and electronic media and in online blogs and advocacy sites. This has stirred widespread concern about the privacy and safety of the civilian population.[32] The potential proliferation of unmanned aircraft operating in civil airspace will be a new phenomenon that will be much more noticeable than a change in the design of flight controls or structural materials in large commercial transports. Privacy concerns are likely driven in part by public concerns about the collection and dissemination of personal data by government and industry in other areas. Getting past the privacy barrier associated with some UAS applications will require an evolution in public perception and trust. For example, UAS should be designed to ensure that data acquired for functions such as communication or navigation cannot be used to violate privacy.

Second, with crewed aircraft, the public will also take note if and when advances in IA systems enable the safe operation of commercial transports with fewer than two pilots, the currently required number. Over time, IA systems could make it possible to transition from two-pilot aircraft to single-pilot operations and, eventually, to fully autonomous aircraft with no pilots on board. To be certain, aircraft manufacturers are not yet ready to unveil pilotless passenger transport aircraft. Even if or when an autonomous commercial passenger and cargo aircraft become available it is likely that public acceptance and trust barriers will be difficult to overcome. A study on public perceptions of unmanned aircraft in 2011 determined that 63 percent of the public would support the operation of autonomous cargo transports without a pilot on board, but just 8 percent would support the operation of autonomous passenger aircraft with no pilot on board. However, in the case of autonomous transports that had a pilot on board who could intervene in an emergency, public support increased substantially: from 63 to 90 percent for cargo flights and from 8 to 77 percent for passenger flights.[33] Perhaps driverless cars can pave the way for trust in pilotless aircraft, despite the fact that achieving high levels of safety in the "driverless car" may be more challenging owing to the dynamics and complexity in the highway environment.

Apart from public concerns about privacy and safety, other social issues are likely to arise. For example, the tendency of people to resist change will likely be a factor. Such resistance will take shape especially when the self-interest of individuals or their organizations are threatened directly or indirectly by the changes likely to arise from the deployment of IA systems.

---

[32] Rachel L. Finn and David Wright, 2012, Unmanned aircraft systems: Surveillance, ethics and privacy in civil applications, *Computer Law and Security Review* 28(2):184-194.

[33] Alice Tam, 2011, *Public Perception of Unmanned Aerial Vehicles,* Purdue University, Publications Department of Aviation Technology, http://docs.lib.purdue.edu/cgi/viewcontent.cgi?article=1002&context=atgrads.

4

Research Agenda

PRIORITIZATION PROCESS

Increasingly autonomous (IA) systems will require research and technology development in many areas. In some cases, that research has been and will continue to be motivated for applications other than civil aviation, and there is no need for aviation-specific research in those areas, which include computing, digital communications, sensors, and open-source hardware and software.

The committee identified a set of high-priority research projects that, if implemented, would address technological issues directly related to the development and application of IA systems in civil aviation. Many of the issues addressed by these projects are related, and many of the research areas included in each project touch on issues of interest to one or more of the other projects. Thus, progress in autonomy research as a whole would be achieved most efficiently by proceeding in a coordinated fashion across multiple areas. Because the identified high-priority projects and the barriers they must overcome are so challenging, they will require government, industry, and academia to perform rigorous and protracted research.

The committee considered a variety of flight crew interaction scenarios ranging from the existing flight deck and remote pilot options that are available today to optionally piloted, unmanned, and unattended (autonomous) operations. Because progress in research tends to be incremental, the committee anticipates a progressive introduction of IA systems into air traffic management (ATM) systems, crewed aircraft, and unmanned aircraft systems (UAS). Similarly, the progressive introduction of UAS operations into shared airspace will also be paced by progress in addressing the high-priority research projects.

Table 4.1 lists the eight high-priority research projects identified by the committee and the barriers addressed by each. A multistep process was used to converge on the eight projects. Committee members were initially briefed by experts in the field and held informal discussions related to these briefings. Each committee member then individually created a list of top research priorities. The compiled list was distributed to the group, organized, and discussed. The committee reached consensus that all of the eight research areas listed in Table 4.1 are a high priority.

The committee could have ranked the projects based on how effectively they improve individual elements of the Observe, Orient, Decide, and Act (OODA) loop or on the extent to which each project contributes to realizing the potential benefits of IA systems described in Chapter 2. However, the contributions of many or most of the research projects are not limited to individual elements of the OODA loop or to the individual benefits that IA systems can provide.

TABLE 4.1 Mapping of High-Priority Research Projects to the Barriers

	High-Priority Research Projects							
	1. Behavior of Adaptive and Nondeterministic Systems	2. Operation Without Continuous Human Oversight	3. Modeling and Simulation	4. Verification, Validation, and Certification	5. Nontraditional Methodologies and Technologies	6. Roles of Personnel and Systems	7. Safety and Efficiency	8. Stakeholder Trust
Technology Barriers								
Communications and data acquisition			●	●		■		
Cyberphysical security			●	●	■	■		●
Diversity of vehicles			●		■	■		
Human–machine integration		■	■				■	
Nondeterministic decision making	■	■	■	■	■			■
Sensing, perception, and cognition		■					■	
System complexity and resilience		●	■	■	■	■		■
Verification and validation	■	●	■	■	■	■		■
Regulation and Certification Barriers								
Airspace access for unmanned aircraft		■	●		●	●	●	●
Certification process	■	■	■	■	■	■	■	■
Equivalent level of safety	■	■	■	■	■	■	■	■
Trust in nondeterministic IA systems	■	●	●	■	■	■		■
Additional Barriers								
Legal issues						■		■
Social issues	●					■		■

NOTE: Black squares indicate that a research project will make a vital contribution to overcoming the corresponding barrier. Circles indicate that the research project will make an important but less vital contribution. Empty cells indicate that the research project will make at best a minor contribution.

Alternatively, the committee could have decided to rank the research projects based on how many barriers they address. However, as the shown in the table, all eight projects address multiple barriers, and the collective research agenda addresses all barriers to some extent. Also, outputs from all of the research projects are ultimately required to allow introducing advanced IA systems into the National Airspace System (NAS).

The committee concluded, accordingly, that level of difficulty and degree of urgency would be the most suitable bases for categorizing the high-priority research projects, because it is those criteria that are most directly related to (1) the extent to which the current state of the art must be advanced; (2) the time and resources needed to make those advances; and (3) the time-phased application of research project results in the overall scheme of developing ever-more-capable IA systems, bringing them to market, and deploying them operationally.

Based on the committee's perception of level of difficulty and urgency, research projects 1 through 4 were determined to be particularly urgent and difficult to accomplish. Projects 5 through 8 are also to be given high priority, but they were determined to be less urgent and/or less difficult. The committee did not rank the research projects within each group of four; instead, they are listed alphabetically.

MOST URGENT AND MOST DIFFICULT RESEARCH PROJECTS

This section describes the four most urgent and most difficult high-priority research projects. The first project addresses the fundamental need to better characterize and bound the behavior of adaptive/nondeterministic systems. The second describes research in system architectures and technologies that would enable increasingly sophisticated IA systems and unmanned aircraft to operate for extended periods of time without real-time human cognizance and control. The third calls for the development of the theoretical basis and methodologies for using modeling and simulation to accelerate the development and maturation of advanced IA systems and aircraft. The fourth calls for research to develop standards and processes for the verification, validation, and certification of IA systems. For each research project, background information is followed by a description of specific research areas and how they would support the development of IA systems for civil aviation.

Behavior of Adaptive/Nondeterministic Systems

Develop Methodologies to Characterize and Bound the Behavior of Adaptive/Nondeterministic Systems over Their Complete Life Cycle

Adaptive/nondeterministic properties will be integral to many advanced IA systems, but they will create challenges for assessing and setting the limits of their resulting behaviors. Advanced IA systems for civil aviation operate in an uncertain environment where physical disturbances, such as wind gusts, are often modeled using probabilistic models. These IA systems may rely on distributed sensor systems that have noise with stochastic properties such as uncertain biases and random drifts over time and under varying environmental conditions. Adaptive/nondeterministic IA systems take advantage of evolving conditions and past experience to improve their performance. As these IA systems take over more functions traditionally performed by humans, there will be a growing need to incorporate autonomous monitoring and other safeguards to ensure continued appropriate operational behavior.

There is a tension between the benefits of incorporating software with adaptive/nondeterministic properties in IA systems and the requirement to test such software for safe and assured operation. Research is needed to develop new methods and tools to address the inherent uncertainties in airspace system operations and thereby assure the safety of complex adaptive/nondeterministic IA systems.

Specific tasks to be carried out by this research project include the following:

- *Develop mathematical models for describing adaptive/nondeterministic processes as applied to humans and machines.* New mathematical models for describing adaptive, nondeterministic, and learning processes as applied to humans and machines are desirable to support the design and testing challenges of IA systems. Current approaches often rely on linearized models with simple normal probability distributions to characterize uncertainty. These simplified representations often do not adequately capture the complexities of IA systems. More complex models create challenges to the design community as their tools commonly assume the same linearized, normally distributed characterizations. The potential behaviors of complex and adaptive/nondeterministic systems operating in environments with many sources of random inputs can quickly overwhelm the conventional tools used in exhaustive deterministic simulation testing. The desired mathematical models would support the goals for effective testing over the wide range of operational conditions as well as support the community that designs algorithms with the desired performance and robustness properties.
- *Develop performance criteria, such as stability, robustness, and resilience, for the analysis and synthesis of adaptive/nondeterministic behaviors.* Performance criteria for such systems, including operational mission envelopes, stability margins, robustness in the face of uncertainty, and resilience to unanticipated events, are needed so that both the design and test communities can effectively account for adaptive/nondeterministic behaviors. These performance criteria would apply and scale to various levels and applications of IA systems, including basic flight performance, onboard aircraft operational decision making, and ATM. Tools to

assess the trade-offs among these criteria are needed along with technologies to remove today's constraints and allow the expansion of performance simultaneously across multiple dimensions.

- *Develop methodologies beyond input-output testing for characterizing the behavior of IA systems.* Testing of aircraft avionics and software is often performed today with a large set of test scripts intended to cover the wide range of expected inputs and anticipated operational configurations. The resulting output response behaviors are evaluated for acceptability. The rapid expansion of computational capabilities from advanced processors and cloud computing, along with applications of experimental techniques, is not sufficient to address the exponential growth in potential input-output behaviors and system configurations for complex IA systems. Therefore new methodologies beyond input-output testing for characterizing the behavior of IA systems are needed.
- *Determine the roles that humans play in limiting the behavior of adaptive/nondeterministic systems and how IA systems can take over those roles.* There is an evolving trade-off in IA systems between the functions performed by humans and by machines. Humans play a key role in monitoring and limiting the behavior of systems as well as in helping IA systems overcome their own limitations, making them more robust in unforeseen situations. Research is needed to identify the roles of humans and to learn how to effectively and safely transition those roles to these systems.

Research into the behavior of adaptive/nondeterministic systems that takes into account the needs of both the design and test communities can have a meaningful effect on advanced IA systems operating in civil airspace.

Operation Without Continuous Human Oversight

Develop the System Architectures and Technologies That Would Enable Increasingly Sophisticated IA Systems and Unmanned Aircraft to Operate for Extended Periods of Time Without Real-Time Human Cognizance and Control

Crewed aircraft have systems with varying levels of automation that operate without continuous human oversight. Even so, pilots are expected to maintain continuous cognizance and control over the aircraft as a whole. Advanced IA systems could allow unmanned aircraft to operate for extended periods of time without human operators to monitor, supervise, and/or directly intervene in the operation of those systems in real time. Eliminating the need for continuous human cognizance and control could achieve multiple objectives:

- *Enhance system performance by better matching the roles of humans and machines to their respective strengths.* For example, repetitive scans of instruments and the environment might be done by IA systems that enable operators to focus more on task- or mission-level decision making. This capability could help maintain the attentiveness of the pilots or operators of crewed aircraft and UAS, and it could also be used to better manage the workload of a single pilot, making the operations safer.
- *Increase the number of aircraft that can be managed by a single human operator and increase the capacity of the civil aviation system.* IA systems capable of safely functioning without continuous human cognizance and control will support intentional and potentially long-term gaps in supervision. Such gaps would need to be acceptable in order to allow single operators to manage multiple aircraft. IA systems could increase capacity and throughput by offering capabilities such as precise and predictable flight plan following, enhanced separation assurance and deconfliction in airspace with positive control,[1] and detect-and-avoid.[2]
- *Enhance the ability of unmanned aircraft to perform dangerous and/or long-endurance missions that are beyond the abilities of humans.* An example of a dangerous mission would be disaster response at the site of a damaged nuclear reactor, where acquiring aerial data (e.g., images and atmospheric chemistry) would

[1] R. Chipalkatty, P. Twu, A. Rahmani, and M. Egerstadt, 2012, Merging and spacing of heterogenerous aircraft in support of NextGen, *Journal of Guidance, Control, and Dynamics* 35(5): 1637-1646.

[2] X. Prats, L. Delgado, J. Ramírez, P. Royo, and E. Pastor, 2012, Requirements, issues, and challenges for sense and avoid in unmanned aircraft systems, *Journal of Aircraft* 49(3): 677-687.

be critical but too hazardous for humans. An example of a long-endurance mission would be polar climate monitoring by recording atmospheric temperature and winds as well as surface ice extent and movement.
- *Facilitate new unmanned aircraft operations where continuous human cognizance and control would not be cost-effective.* For example, unmanned aircraft supporting large-scale agriculture operations would require long-range flights that go beyond the line of sight of the operator and that collect data overnight (when other farming operations are shut down). In other markets, a host of small unmanned aircraft applications could become profitable if they did not need to be continuously supervised by a human operator.

Specific tasks to be carried out by this research project include the following:

- *Investigate human roles, including temporal requirements for supervision, as a function of the mission, capabilities, and limitations of IA systems.* This includes a variety of IA system operational scenarios ranging from full-time human supervision to potentially no supervision. Risk analyses that consider mission, vehicle type, and environment (airspace and ground) could identify operational scenarios in which continuous cognizance and control are not required. The requirements for supervision might be periodic (e.g., an operator might be required to regain cognizance and control once every 15 minutes), event-based (e.g., an operator is alerted whenever a problem is detected by the IA system), or both. It will be important to understand how IA can appropriately alert operators and help them regain situational awareness when requests for operator input are event-based.
- *Develop IA systems that respond safely to the degradation or failure of aircraft systems.* Onboard hardware and software failures can pose a serious risk, particularly when safety-critical systems provide erroneous data or become unavailable. Performance degradation—due, for example, to structural damage—also increases risk. Autopilots incorporated in current flight management systems disengage when an exception condition is triggered by a degradation or failure. This sudden and potentially unexpected event can place sudden demands on operators, in terms of workload, skill, and decision making, that are beyond reasonable levels. IA systems can address many of the challenges associated with degradation or failure conditions. For example, IA systems could recognize the extent of a failure, determine a course of action, and execute a safe recovery.
- *Develop IA systems to identify and mitigate high-risk situations induced by the mission, the environment, or other elements of the NAS.* Examples include loss of separation from other aircraft or terrain and flight into adverse weather conditions. Mitigation could involve warning and informing human supervisors about risk factors as well as engaging IA systems to rapidly mitigate the risk, for example, by recovering stable flight in loss-of-control situations. Key IA technologies contributing to this research include diagnostic, prognostic, and decision-making systems that may be adaptive and/or nondeterministic. Advanced sensing (e.g., onboard or ground-based radar), database access (e.g., weather), and effective communication with operators and ATM systems could enhance onboard IA capabilities.
- *Develop detect-and-avoid IA systems that do not need continuous human oversight.* IA detect-and-avoid systems can reduce the possibility of collision or loss of separation for crewed and unmanned aircraft. The ATM system, supplemented by systems such as the Traffic Alert and Collision Avoidance System (TCAS)[3] and Automatic Dependent Surveillance-Broadcast (ADS-B) provide human operators with data that may be critical in an emergency to assure separation and to maneuver aircraft to avoid collision if separation is lost. Relying on detect-and-avoid systems that operate without human supervision would require that the IA system be sufficiently reliable and secure to function correctly in all shared airspace. Redundancy, for example, through ADS-B and radar systems, will be key to ensuring sufficient reliability.[4] Aircraft performance and maneuverability differences will need to be accommodated in detect-and-avoid

[3] TCAS is an airborne system that alerts pilots to conflicting air traffic and provides guidance on maneuvers to avoid collisions.

[4] S. Cook, A. Lacher, D. Maroney, and A. Zeitlin, 2012, UAS sense and avoid development—The challenges of technology, standards, and certification, *Proceedings of the AIAA Aerospace Sciences Meeting,* January 2012, Nashville, Tenn., Paper AIAA-2012-0959, Reston, Va.: American Institute of Aeronautics and Astronautics.

solutions. For example, detect-and-avoid systems should be able to handle deconfliction scenarios with other aircraft that are not equipped with ADS-B and with aircraft whose pilots who respond inappropriately to traffic alerts.

- *Investigate airspace structures that could support UAS operations in confined or pre-approved operating areas using methods such as geofencing.* Aircraft performing missions such as agricultural or pipeline inspections typically operate over unpopulated areas at low altitudes. Such missions pose little risk to people or property on the ground but may pose risk to other aircraft. IA systems for geofencing can ensure a vehicle operates within a given airspace volume designated by constraints on altitude ceiling, latitude, longitude, and, in some cases, time. Such a geofencing system will likely need to be matured and certified even if other systems on the geofenced aircraft are not. For geofencing to be useful, IA systems would also need to support dynamic allocation of (low-altitude) airspace to geofenced flight operations. Such IA systems can reduce collision risk for both geofenced and legacy aircraft.

The Department of Defense (DOD) has invested substantial effort in technologies related to this research project. Areas of interest include autopilots for fighters that can execute and recover from maneuvers that may induce pilots to lose consciousness, unmanned aircraft that can continue a mission or return home following loss of the communications link to ground operators, and UAS ground stations from which one operator is responsible for multiple unmanned aircraft. NASA has studied IA systems for airspace management and loss-of-control prevention and has successfully demonstrated that unattended operation of orbiting spacecraft can reduce mission costs. Numerous researchers and commercial vendors are developing and demonstrating IA systems with the potential to eliminate the requirement for continuous human cognizance and control for robotic system applications, including small unmanned aircraft and automotive applications. DOD, NASA, and industry have the expertise to pursue efforts in this research direction. The Federal Aviation Administration (FAA), which has historically presumed continuous human cognizance and control, would need to expand its expertise and engage the community in this research task to efficiently establish standards and policies to authorize the operation of unmanned aircraft without continuous human oversight.

Modeling and Simulation

Develop the Theoretical Basis and Methodologies for Using Modeling and Simulation to Accelerate the Development and Maturation of Advanced IA Systems and Aircraft

Modeling and simulation capabilities will play an important role in the development, implementation, and evolution of IA systems in civil aviation because they enable researchers, designers, regulators, and operators to understand how something performs without real-life testing. Civil aviation has leveraged modeling and simulation throughout its history. Examples include computational fluid dynamic models and human-in-the-loop simulations to test operational concepts and decision support tools. Modeling and simulation enable insights into system performance without necessarily incurring the expense and risk associated with actual operations. Furthermore, computer simulations may be able to test the performance of some IA systems in literally millions of scenarios in a short time to produce a statistical basis for determining safety risks and establishing the confidence of IA system performance.

Researchers and designers are likely to make use of modeling and simulation capabilities to evaluate design alternatives. Developers of IA systems will be able to train adaptive (i.e., learning) algorithms through repeated simulated operations. Such capabilities could also be used to train human operators.

Modeling and simulation capabilities will be a tool leveraged in addressing several of the other projects identified in this report, such as characterizing the behavior of adaptive/nondeterministic systems; generating the data and artifacts needed for VV&C; generating data for determining how IA systems could enhance the safety and efficiency of the civil aviation system; providing a mechanism to evaluate stakeholder roles and responsibilities and human–machine interfaces of IA systems; and producing data to inform rulemaking and policy changes.

Specific tasks to be carried out by this research project include the following:

- *Develop theories and methodologies that will enable modeling and simulation to serve as embedded components within adaptive/nondeterministic systems.* Modeling and simulation capabilities are likely to be embedded as parts of adaptive/nondeterministic IA systems. An introspective system would be capable of passing judgment on its own performance. This would require the ability to model its own behavior and predict the outcome in a variety of environmental circumstances in which it is intended to act.
- *Develop theories and methodologies for using modeling and simulation to coach adaptive IA systems and human operators during training exercises.* Many IA systems will have adaptive or learning algorithms as components of the system. Adaptive algorithms evolve internal parameters as a function of historical performance. Modeling and simulation could be used to help evolve these internal parameters by creating a virtual operating environment enabling literally millions of scenarios to be experienced by the IA systems and their associated adaptive algorithms. The simulated scenarios and resulting performance would serve as feedback to the adaptive algorithm, enabling it to learn and improve without actual operations.
- *Develop theories and methodologies for using modeling and simulation to create trust and confidence in the performance of IA systems.* Modeling and simulation capabilities could be used to generate performance data and demonstrate the behavior of IA systems over a wide range of operational scenarios. Evaluations will need to be conducted in a manner that models the environmental circumstances in which the IA system is intended to operate. These performance data will be used to help create trust and confidence in the performance of IA systems. In some instances, modeling and simulation of virtual environments could be linked with IA systems operating in the real world.
- *Develop theories and methodologies for using modeling and simulation to assist with accident and incident investigations associated with IA systems.* Modeling and simulation capabilities can also be a forensic tool for analysts trying to understand the behavior of IA systems when investigating the factors that may have contributed to an accident, incident, or other anomaly. To assist analysts, IA systems will need to be instrumented to collect operational data for subsequent analysis. A functional equivalent of flight data recorders may need to be devised to understand what the IA system might have been "thinking" when making certain decisions or taking specific actions.
- *Develop theories and methodologies for using modeling and simulation to assess the robustness and resiliency of IA systems to intentional and unintentional cybersecurity vulnerabilities.* By their very nature, IA systems are complex cybersystems that may be vulnerable to intentional cyberattacks or unintentional combinations of circumstances. Modeling and simulation can create virtual environments to evaluate how well the system will continue to perform in a variety of circumstances.
- *Develop theories and methodologies for using modeling and simulation to perform comparative safety risk analyses of IA systems.* IA systems will be used to supplement or replace existing mechanisms in the current aviation system. Their use will introduce new hazards even as they eliminate or mitigate existing hazards. Modeling and simulation tools will be needed to assist the aviation community in comparing the risks of continuing to operate with the legacy systems on the one hand and employing the IA systems on the other.
- *Create and regularly update standardized interfaces and processes for developing modeling and simulation components for eventual integration.* Monolithic modeling and simulation efforts intended to develop capabilities that can "do it all" and answer any and all questions tend to be ineffective owing to limitations on access and availability; the higher cost of creating, employing, and maintaining them; the complexity of application, which constrains their use; and the centralization of development risks. Rather, the committee envisions the creation of a distributed suite of modeling and simulation modules developed by disparate organizations with the ability to be interconnected or networked as appropriate. To ensure that these systems can be interconnected, modeling and simulation capabilities will need to meet community standards, which themselves will need to be established and periodically reviewed and updated.
- *Develop standardized modules for common elements of the future system, such as aircraft performance, airspace, environmental circumstances, and human performance.* To reduce the cost of research, development, and analysis, reuse of common modeling elements for the future system could be encouraged by

developing them in a manner that allows them to be integrated through established standards. These elements also need to be readily accessible. A mechanism could be established that would allow organizations involved with researching and developing IA systems to share modeling components. This would facilitate reuse of previously developed components, saving time and resources.
- *Develop standards and methodologies for accrediting IA models and simulations.* Given the importance of modeling and simulation capabilities to the creation, evaluation, and evolution of IA systems, mechanisms will be needed to ensure that these capabilities perform as intended. Just as computational fluid dynamic models are validated with wind tunnel tests, and wind tunnels are validated with flight tests, a means to validate the modeling and simulation capabilities for IA systems will be needed.

Verification, Validation, and Certification

Develop Standards and Processes for the Verification, Validation, and Certification of IA Systems and Determine Their Implications for Design

The high levels of safety achieved in the operation of the NAS largely reflect the formal requirements imposed by the FAA for VV&C of hardware and software and the certification of personnel as a condition for entry into the system. These processes have evolved over many decades and represent the cumulative experience of all elements of civil aviation—manufacturers, regulators, pilots, controllers, other operators—in the operation of that system. Although viewed by some as unnecessarily cumbersome and expensive, VV&C processes are critical to the continued safe operation of the NAS.

Existing VV&C standards and processes are inadequate for the development and regulation of advanced IA systems with attributes such as nondeterministic outcomes, emergent behaviors, vehicle teaming, and software and/or system changes arising from machine learning. Current practice involves a comprehensive examination of all system states in all known conditions. This approach is neither practical nor feasible from the standpoints of timeliness and affordability for adaptive/nondeterministic IA systems. Fixed-logic models that contain closed-world assumptions, where all system interactions are predetermined and known, are unable to account for unplanned events falling outside the conditions that were considered in the reasoning process. Given the inevitable uncertainty of the real-world environment, current safety methods rely heavily on the human operator to mitigate system anomalies and unanticipated events. With IA systems, a portion of this responsibility would shift to the IA system.

Specific tasks to be carried out by this research project include the following:

- *Characterize and define requirements for intelligent software and systems.* The issue of trust resides in trust verification and trust validation[5] for both the software-based "intelligence" and the systems in which it is embedded. The ability of such systems to function acceptably in real-world conditions requires the development of trust in the IA system. Thus, it will be important to identify and characterize the attributes, characteristics, and metrics required to enable trust verification of the intelligent software used by advanced IA systems through interaction with testers and/or examiners in a manner analogous to how human performance is assessed in civil airspace today.
- *Improve the fidelity of the VV&C test environment.* To ensure that the VV&C processes for IA systems can substantiate the intended requirements to effectively manage unplanned events, novel new verification and validation (V&V) tools, criteria, and evaluation methodologies need to be developed, including representative test cases and trust metrics that fully identify and characterize the capabilities of IA systems. This will require significant research in V&V methodologies, focusing on the full range of uncertain behaviors and environmental conditions expected to be encountered in the airspace system. Stress testing of this underlying IA system logic will assess the extent to which nominal, off-nominal, and even unanticipated

[5] Trust verification is the process of assuring that the intelligent software of an IA system is trusted to perform as expected in the tasks, missions, and environments where it is intended to be utilized. Trust validation is the process of assuring that the system as a whole (hardware, software, and human operator) is trusted to perform as expected in the tasks, missions, and environments where it is intended to be utilized.

cases can be accommodated. Confronting the full IA system with as-yet-to-be-experienced situations in realistic scenarios (e.g., detecting taxiing aircraft on a runway during final approach) is key to identifying shortcomings in the capabilities of an advanced IA system. Using objective measures within this evaluation process is intended to lead to greater predictability and trust in the logic of the IA system. To maintain an equivalent level of safety, it is incumbent on the developers to assure that IA systems have undergone these rigorous V&V processes, including safety risk assessments. Mission diversity and complexity create design challenges, because the decision making, sensing, and experience level required may vary significantly from mission to mission. This complexity and diversity will require the system to be completely reconfigurable to ensure safe and successful mission execution, creating challenges for the VV&C process. Elements of this research include the following:

—Integration of modeled and actual subsystems, such as sensors, reasoning algorithms, hardware components, and human operators.

—Simulation of low- and high-fidelity tests.

—Research into VV&C processes for interoperability frameworks that allow rapidly configurable trust verification and trust validation.

—Investigation of the impacts of system- and mission-level disruptions in the human-hardware-software system, such as the loss of Global Positioning System inputs, control links, and sensors.

- *Develop, assess, and propose new certification standards.* In addition to the development of rigorous new V&V processes, new certification standards and practices will need to be written with guidance for the regulatory authorities on understanding how IA systems function and what it will take to establish a means of compliance with the yet-to-be-established certification processes for advanced IA systems. Once certified, it may be necessary to continually monitor IA system compliance with requirements and to assess the performance of nondeterministic IA systems as they learn and adapt. Elements of this research include the following:

  —Establishment of criteria and standards uniquely focused on the emergent behaviors of nondeterministic IA systems and approaches for assuring continual adherence to these standards.

  —Identification of critical pass-fail boundaries that can be used to assure safe operation of IA systems for different mission sets in civil airspace.

  —Exploration of novel certification approaches that would take place concurrently with the design and development of advanced IA systems, as well as novel techniques to enable continuous retest of nondeterministic IA systems.

- *Define new design requirements and methodologies for IA systems.* In the course of setting a clear path to certification and eventual deployment, unique design requirements could be created for IA systems so that they can interoperate with other elements of the NAS. Emergent behaviors could lead to fundamental changes to the design methodologies and design characteristics required of the IA systems. For example, to understand and accept an IA system's actions as reasonable and predictable, it will be critical to understand the underlying logic of the decision-making process. Understanding the logic enables the stakeholders within the airspace to infer intent, which adds to the system's predictability, credibility, and trust. New design frameworks need to address mission diversity and complexity while simultaneously simplifying the VV&C process for IA systems. Elements of this research include the following:

  —Investigation of new and agile system design approaches and methodologies to enable integration into civil airspace and trust in IA systems.

  —Enabling transparency for the rationale and decision making of an IA system for given situations.

  —Development of high-level intelligent software tools to reveal an IA system's underlying logic in situ and, if required, alter it.

  —Creation of design frameworks (interoperability, complex networks, plug and play, and so on) that address mission complexity and diversity while simplifying VV&C processes.

- *Understand the impact that airspace system complexity has on IA system design and on VV&C.* The effectiveness of IA systems will depend significantly on how they interact with the NAS. Real-world operational issues such as system latency, degraded sensor performance, and intermittency of communications will be amplified by mission and environmental conditions. Elements of this research include the following:

—Developing design methodologies that effectively and comprehensively address IA system requirements over a broad range of intended operating conditions, encompassing geographical complexity (e.g., rural versus urban), mission complexity (e.g., single vehicles used for agriculture versus multiple high-precision aircraft), and environmental changes (e.g., weather anomalies).

—Developing the ability to leverage the maturation of IA systems over time with regard to airspace complexity, such as acquisition of experiential knowledge of operations at diverse classes of civil airspace and analysis of incidents, accidents, and other anomalies involving IA systems.

- *Develop VV&C methods for products created using nontraditional methodologies and technologies.* Some new civil aviation products will be developed using nontraditional methodologies and technologies (see the discussion below of the research project of the same name). The deployment and operation of products such as consumer electronics provide large amounts of user data that can be used for validation and verification when the technology is applied to civil aviation. Although the information generated from deployment in one application or industry is not complete, it could nevertheless be considered in the overall certification process. Furthermore, unconventional techniques such as crowd sourcing[6] can increase levels and volumes of testing, including the diversity and uniqueness of test scenarios that could not or would not be achieved by traditional methods. These or similar test techniques might be used in conjunction with more traditional test methodologies to accelerate the acceptance process.

ADDITIONAL HIGH-PRIORITY RESEARCH PROJECTS

This section describes four additional high-priority research projects. They are distinguished from the four already discussed by their lower degree of urgency and/or difficulty. The first of these final four research projects would develop methodologies for accepting technologies not traditionally used in civil aviation in IA systems. The second focuses on how the roles of key personnel and systems, as well as human–machine interfaces, should evolve to enable the operation of advanced IA systems. The third would analyze how IA systems could enhance the safety and efficiency of civil aviation. The fourth would investigate processes to engender broad stakeholder trust in IA systems for civil aviation. For each research project, background information is followed by a description of specific research areas and how they would support the development of IA systems for civil aviation.

Nontraditional Methodologies and Technologies

Develop Methodologies for Accepting Technologies Not Traditionally Used in Civil Aviation (e.g., Open-Source Software and Consumer Electronic Products) in IA Systems

Many key elements of IA systems have their roots in the commercial world of information technology, which has long tended toward increasing capability at decreasing cost. The same cannot be said of aircraft avionics. Civil aviation in general and IA systems in particular can benefit directly and indirectly from a broad variety of technologies developed through other nontraditional methods in areas that include low-cost, high-capability computing systems; digital communications; miniaturized sensor technology; position, navigation, and timing systems; open-source hardware and software; mobile communications and computing devices; and small unmanned aircraft developed for personal or hobbyist use. Furthermore, IA technologies have the potential to spawn new nontraditional uses, applications, and industries. Methodologies are needed that would enable IA systems that incorporate these nontraditional technologies to be integrated in the NAS.

Enabling the acceptance of nontraditional technologies, methods, and uses can have a variety of immediate and future benefits. For example, the large and growing community of hobbyists who work on UAS is currently

[6] Crowd sourcing relies on the general public to contribute funds or labor to a project, most commonly via the Internet and without compensation.

building and operating such devices for personal use. These could be but often are not operated within the FAA's very narrow guidelines for recreational flight.[7] New acceptance methods would provide a pathway to the commercialization and legal use of personal UAS, which could in turn clear the way for many new UAS applications. In addition, new methodologies will enable the incorporation of IA capabilities developed for and currently used in other industries by means of diverse and often very different V&V processes. For example, the automotive industry is deploying IA systems, and the large number of automobiles on the road means that these systems will undergo extensive field testing and evaluation. In addition, accepting nontraditional technologies could exploit some rapidly evolving and proliferating emerging technologies and systems and reap their economic benefit. For example, personal computing devices such as tablet computers, whose capabilities have been advancing at a phenomenal rate, are now often used as a pilot aid by private and commercial pilots.

Specific tasks to be carried out by this research project include the following:

- *Develop modular architectures and protocols that support the use of open-source products for non-safety-critical applications.* Open-source software and products are prominent in several technology areas that could support IA systems, especially in the areas of mobile computing and network communications. The openness of the product development process increases the potential number of developers and accelerates the pace of innovation and change. The pace at which advances occur in this arena is at odds with the time needed for traditional certification processes and their complexity. New modular architectures and protocols are needed to accommodate this pace. Non-safety-critical system elements and applications that inherently involve little risk, such as the operation of small unmanned aircraft over fields on large farms, are natural candidates for focusing initial efforts.
- *Develop and mature nontraditional software languages for IA applications.* As intelligent machines evolve, the enabling software will require new attributes unavailable from today's deterministic code. This software will require an architectural underpinning that (1) permits integration of multiple tasks, often conducted concurrently; (2) is autonomously reconfigurable as mission and environmental circumstances warrant; (3) is scalable to encompass large, complex systems functionality; and (4) is verifiable and trustworthy in its domain of applicability.
- *Develop paths for migrating open-source, intelligent software to safety-critical applications and unrestricted flight operations.* Additional processes could be developed for migrating open-source software to safety-critical applications and unrestricted flight operations. Open-source development does not imply simple or unsafe functionality, as evidenced by the Linux computer operating system, which is used in a large variety of mainframes and servers, personal computers, and consumer devices. Approaches such as software-based "functional redundancy" could be explored to increase overall system reliability as an adjunct to traditional hardware redundancy.
- *Define new operational categories that would enable or accelerate experimentation, flight testing, and deployment of nontraditional technologies.* In order to provide the regulatory flexibility to accept nontraditional technology into civil aviation, new operational categories could be developed that specifically focus on the advancement of IA capabilities, recognizing overall system safety, risk, performance, and so on, as the guiding factors rather than component development process or pedigree. These categories should include operations in confined or nonnavigable airspace. This type of mechanism could accelerate maturation of nontraditional applications and missions, such as beyond-line-of-sight precision agriculture.

The final phase of this research project would be accomplished by the VV&C research project as it develops certification standards.

[7] FAA, 1981, Advisory Circular AC 91-57, June 9.

Roles of Personnel and Systems

Determine How the Roles of Key Personnel and Systems, As Well As Related Human–Machine Interfaces, Should Evolve to Enable the Operation of Advanced IA Systems

Effectively integrating humans and machines in the civil aviation system has been a high priority and sometimes an elusive design challenge for decades. Human–machine integration may become an even greater challenge with the advent of advanced IA systems. Although the reliance on high levels of automation in crewed aircraft has increased the overall levels of system safety, persistent and seemingly intractable issues arise in the context of incidents and accidents.

The very nature of IA systems dictates the need for new methods to interface and interact with their human operators and other counterparts. The prospect of shared human–machine intelligence presents daunting challenges ranging from technical to infrastructural to ethical. The emergence of advanced IA capabilities in civil aviation will also require changes to the definition and management of operations within the NAS. The role of key personnel and systems will evolve in response to these changes. In some instances new roles may be added while others become less significant or cease to add value. Advanced IA systems may reduce the number of air traffic controllers, pilots, and other human operators needed by the NAS, while the need for IA system management and supervision might actually increase. Some of today's active control roles will transition to monitoring functions and may ultimately become extinct. In any case, the roles of these key personnel will have to be revisited in relation to their contribution to and impact on the NAS and the requirements imposed by the adoption of emerging technologies.

Much of the research to be conducted revolves around the roles that human beings currently play and the roles that they should play in interacting with intelligent machines. There is no simple answer. In fact, the answer depends on the nature of the system intelligence and can be expected to change over time as technologies and machine capabilities evolve. As an example, in contrast to the nearly continuous interaction that a pilot has with a highly automated flight control system, which is now the case on commercial transports, interactions between an intelligent, autonomous mission management system and its human supervisor could be highly intermittent. As it now stands, the flight control system executes its decision cycle an order of magnitude (or more) faster than the pilot, who issues input commands via manual ergonomic means (e.g., stick or yoke, rudder pedals, throttle) and specifies the coordinates of way points. In the not-too-distant future, advanced IA systems may be able to assume primary responsibility for functions such as hazardous weather avoidance.

More capable IA systems will offer opportunities for reallocation of functions within the integrated human–machine system. Where practical, humans and machines could each be allocated the functions for which they are best suited while maintaining a secondary capability in functions assigned to the other one. Individual and systemic resilience would be enhanced by maintaining this degree of dual functionality. However, many of the functional issues ascribed to today's automated systems can be traced to some combination of overlapping capabilities (manual operation vs. automated function), insufficient integration of these functions, and deficiencies in system design and/or training that impede the smooth transition from one mode to the other under some operating conditions.

The allocation of functions between humans and machines could be dynamic, allowing role changes based on evolving factors such as fatigue, risk, and surprise. Regardless, new methods are and will be needed for advanced IA systems to assure that human operators smoothly transition to a more proactive command-and-control mode, as needed. As the intelligence of IA systems advances, new opportunities for improved decision making and task execution by the human-machine system will arise, and additional research will be needed to determine how best to take advantage of these opportunities.

The human–system integration problem becomes even more complex with the hierarchical control imposed by flight in managed airspace. This system-of-systems problem, involving onboard pilots, remote UAS operators, intelligent machines, air traffic controllers, and an intelligent ATM network, is perhaps the most daunting human–system integration challenge that civil aviation will face, and overcoming it will require research in relevant technical disciplines and social sciences.

Specific tasks to be carried out by this research project include the following:

- *Develop human–machine interface tools and methodologies to support operation of advanced IA systems during normal and atypical operations.* As IA systems evolve, the need for more robust human–machine interfaces will grow. This research will involve understanding the various options for dynamic human–machine integration, including their respective strengths and weaknesses. Both machines and humans will need tools and methodologies to execute the OODA loop[8] in harmonious fashion for mission accomplishment and safe operation. It may be necessary for machines and humans to share input data, to understand each other's perspectives, and to be aware of the logic behind their respective decisions. If they make different decisions on how to proceed, some means for resolving the differences will be needed. Alternatively, it may be possible to improve the decision rationale used by humans and machines by selectively applying the best elements of each. Unambiguous communication between humans and machines is also important. The use of traditional displays could be a starting point for more robust, meaningful ways to share information and insight between humans and machines. This could include voice, tactile responses, gesture recognition, and other novel means of communicating.
- *Develop tools and methodologies to ensure effective communication among IA systems and other elements of the NAS.* At present, communication between a single air traffic controller and multiple aircraft on one frequency occasionally results in the omission of critical information, overlapping transmissions that block reception, and limited opportunity for clarification or questioning of transmitted information. Data link communication alleviates some of these concerns, but it presents challenges of latency and commonality of terminology in free text communications. Advanced IA aircraft will require a new paradigm to overcome frailties in connectivity and foster concise, unambiguous communications. Research is required to identify and develop effective means of communication between air- and ground-based IA systems. The impact of flight path deviations due to weather, turbulence, traffic congestion, and so on would need to be communicated, along with viable, real-time solutions.
- *Define the rationale and criteria for assigning roles to key personnel and IA systems and assessing their ability to perform those roles under realistic operating conditions.* Key personnel and IA systems will need to operate as an integrated system to operate safely and efficiently in the increasingly complex and congested NAS. Roles of key personnel would need to be identified and aligned with the degree of autonomy and then adjusted over time as the technology and IA capabilities evolve. For example, the roles of ground-based UAS operators and air traffic controllers may change dramatically if an advanced IA ATM network can issue commands directly to onboard autonomous mission management systems. The role of UAS operators could be even more dramatically different if they are able to simultaneously operate multiple aircraft as a constellation of unmanned aircraft. IA systems may need to monitor human workload and performance to dynamically shift allocated tasks for optimum human–machine or network performance. Conversely, humans may need to monitor IA systems to determine and adjust their ability to execute roles.
- *Develop intuitive human–machine integration technologies to support real-time decision making, particularly in high-stress, dynamic situations.* Such situations impose hefty workloads on humans, often reducing their physical, cognitive, or sensory performance. Advanced IA systems provide an opportunity to more effectively manage these workloads, allowing for improved decision making and task execution. As IA capabilities advance, high-stress, dynamic situations may be an area where machine intelligence and decision making could augment or even supplant human intelligence.
- *Develop methods and technologies to enable situational awareness that supports the integration of IA systems.* Awareness of situations such as potentially conflicting decisions, environmental conditions, and the status of mission execution is essential to mission safety and success. The development of IA systems with improved sensing technologies could greatly enhance situational awareness for both the human and the IA system. Research into human and machine perception and the resulting situational awareness could lead to safer and more efficient operations. The research could determine what sensing technologies need

[8] The OODA loop (Observe, Orient, Decide, Act) is discussed in Chapter 1.

to be developed to enable the required situational awareness. It could also explore sensor options that can reproduce the value added from the pilot's seat-of-the-pants experience for both the onboard IA system and the remote operator, if warranted. Sensor research could also explore the cuing required for off-board operators to remain effective during remote operating conditions.

Safety and Efficiency

Determine How IA Systems Could Enhance the Safety and Efficiency of Civil Aviation

As with other new technologies, poorly implemented IA systems could put at risk the high levels of efficiency and safety that are the hallmarks of civil aviation, particularly for commercial air transportation. However, done properly, advances in IA systems could enhance both the safety and the efficiency of civil aviation. For example, IA systems have the potential to reduce reaction times in safety-critical situations, especially in circumstances that today are encumbered by the requirement for human-to-human interactions. The ability of IA capabilities to rapidly cue operators or potentially render a fully autonomous response in safety-critical situations could improve both safety and efficiency. IA systems could substantially reduce the occurrence of these classes of accidents, which are typically ascribed to operator error. These benefits could be of particular value in those segments of civil aviation, such as general aviation and medical evacuation helicopters, that have accident rates much higher than commercial air transports.

Whether on board an aircraft or in ATM centers, IA systems also have the potential to reduce manpower requirements, thereby increasing the efficiency of operations and reducing operational costs.

Where IA systems make it possible for small unmanned aircraft to replace crewed aircraft, the risks to persons and property on the ground in the event of an accident could be greatly reduced owing to the reduced damage footprint in those instances, and the risk to air crew is eliminated entirely.

Specific tasks to be carried out by this research project include the following:

- *Analyze accident and incident records to determine where IA systems may have prevented or mitigated the severity of specific accidents or classes of accidents.* A wealth of historic accident and incident data, including causal information, is archived and available for study. Research could focus on understanding if root causes could have been countered by an IA system and, if not, if an IA system could have mitigated or prevented the consequences of the incident or accident. This research could also explore other methods for avoidance and mitigation to achieve balance and reduce any bias toward IA approaches. Such research could be broad based, assessing the risk to people and property posed by aircraft of widely varying sizes, weights, performance characteristics, and missions, and by various types of operations (ramp, ground, transport, etc.), affording a comprehensive perspective on the value proposition for IA systems and capabilities in this regard.
- *Develop and analytically test methodologies to determine how the introduction of IA systems in flight operations, ramp operations by aircraft and ground support equipment, ATM systems, airline operation control centers, and so on might improve safety and efficiency.* This task would develop quantifiable methods of assessing the value of IA capabilities for various elements of the NAS. Methods of interest could include measures of merit and key performance parameters that map onto safety and efficiency metrics. For example, this research element could develop guidelines and methods to allow current safety methodologies, including nonpunitive incident reporting programs such as the Flight Operations Quality Assurance program and the FAA's Aviation Safety Information Analysis and Sharing system, to help assess the impact of operational IA systems on safety and efficiency. The analysis tools would be able to model the effects, both positive and negative, of IA systems on other portions of the NAS. This research could also determine the training, education, and experience requirements for systems safety personnel, such as accident investigators and airworthiness and operations inspectors, who are involved with IA systems operations.
- *Investigate airspace structures and operating procedures to ensure safe and efficient operations of legacy and IA systems in the NAS.* The introduction of IA systems will increase the complexity and diversity of the NAS, which is already struggling with legacy systems and high traffic volumes. The introduction of

large numbers of UAS by hobbyists, farmers, law enforcement agencies, and others would bring new complexities and place additional stress on the NAS. Airspace partitioning is one of the fundamental tools of risk management used in the current aviation system. Various categories of airspace are defined based on a great variety of factors, including weather, terrain, time of day (day vs. night), traffic type, traffic density, aircraft equipage and performance, and crew qualifications. These factors determine which operations are permitted in a particular region of airspace at any given time. Advanced IA systems might present an opportunity and/or an incentive to redesign some regions of civil airspace, operating conditions, and rules. For example, IA systems might make it feasible and beneficial to dynamically reconfigure some airspace to improve the safety and efficiency of airspace for selected applications.

Stakeholder Trust

Develop Processes to Engender Broad Stakeholder Trust in IA Systems in the Civil Aviation System

IA systems can fundamentally change the relationship between people and technology. Although increasingly used as an engineering term in the context of software and security assurance, trust is a social term and becomes increasingly relevant to human–machine interactions when technological complexity exceeds human capacity to fully understand its behavior. Trust is not a trait of the system; it is the system status in the mind of human beings based on their perception of and experience with the system. Trust concerns the attitude that a person or technology will help achieve specific goals in a situation characterized by uncertainty and vulnerability.[9] It is the perception of trustworthiness that influences how people respond to a system.

Trust in technology is different from trust between people, but there are similarities. Increasing levels of autonomy can blur the distinction from trust in people and trust in technology, because the advanced IA systems might behave in ways that are hard to differentiate from the ways of a person. The basis of this trust includes experience with the system, an understanding of the underlying process, and knowledge of its purpose. Trust can also be based on the recommendation of third parties based on their experience or on their formal certification of a system, or it can reflect informal information sharing that defines the reputation of the trustee. Importantly, the trust engendered by a technology depends in part on an analytic assessment of its capabilities, but it also depends on an intuitive assessment of its behavior.

Fostering an appropriate level of trust in IA systems is critical to overcoming barriers to adoption and acceptance by the broad stakeholder community, which includes operators, supervisors, acquirers, patrons, regulators, designers, insurers, the operating community, and the general public. An individual operator who interacts with an IA system but harbors unwarranted skepticism about the reliability and performance of autonomous systems in general might fail to capitalize on their capabilities and instead rely on less efficient approaches. At the societal level, unwarranted skepticism might cause patrons to seek other services, and the public might even push for legislation that unnecessarily limits the pace of innovation. On the other hand, excessive trust could lead to failures to adequately supervise imperfect IA systems, potentially resulting in accidents and a loss of trust that could be very difficult to regain. Trust influences how people respond to IA systems across various organizational levels, circumstances, and timescales.

Although closely related to V&V and certification, trust warrants attention as a distinct research topic because formal certification does not guarantee trust and adoption. A significant component of this project could be cybersecurity and related issues. Trustworthiness depends on the intent of designers and on the degree to which the design prevents both inadvertent and intentional corruption of system data and processes.

Specific tasks to be carried out by this research project include the following:

- *Identify the objective attributes of trustworthiness and develop measures of trust that can be tailored to a range of applications, circumstances, and relevant stakeholders.* A central challenge in supporting effective

[9] J.D. Lee and K.A. See, 2004, Trust in automation: Designing for appropriate reliance, *Human Factors* 46(1): 50-80.

use of IA systems concerns the need to calibrate the level of trust with the trustworthiness of the system. This would require a metric for trustworthiness of the system, which, like risk, might not be determined by a technical assessment of the IA system.[10] It would also require measures of stakeholder trust in the system that are comparable to the measures of trustworthiness. This would identify situations when people might neglect the system just when it actually requires human cognizance and control. Stakeholder trust is relevant for a broad range of stakeholders and situations, but its measurement and aggregation present a substantial challenge.

- *Develop a systematic methodology for introducing IA system functionality that matches authority and responsibility with earned levels of trust.* The authority of an IA system should be commensurate with its trustworthiness and the degree of trust it earns from the stakeholders. This calls for introducing increased autonomy in a way that both demonstrates trustworthiness and builds trust. Possible methods include graduated authority, in which less and less attention is paid to supervising the system as it demonstrates increasing competence. Another possibility is declarations of third-party trust, whereby developers, other users, and/or an agency testify to the capabilities and characteristics of the system. Because the use of advanced IA systems is likely to grow over time, they would need to be continually assessed in the face of changing technology and situations.
- *Determine the way in which trust-related information is communicated.* The diversity of stakeholders, applications, and circumstances makes it critical to describe the behavior of IA systems in terms that can be understood. Because trust is based on both analysis and intuition, specifying the level at which information is to be delivered is not sufficient to engender trust; in addition, the form in which the information is presented must allow stakeholders to understand what the system is doing and why. In some cases the passive receipt of information will be sufficient; in other cases it may be important to allow users to influence the system's response with specific commands or by manipulating external stimuli.
- *Develop approaches for establishing trust in IA systems.* Trust plays an important role in relationships among individuals, teams, and organizations. Trust also affects the way that people interact with and use conventional technology. Research into both of these areas will be relevant to research into the role that trust plays in human interactions with IA systems. Making good use of this past research, however, will require a better understanding of the differences and similarities between trust in people and machines and how those similarities and differences relate to the specific characteristics of IA systems.

COORDINATION OF RESEARCH AND DEVELOPMENT

All of the research projects described above could and should be addressed by partnerships involving multiple organizations in the federal government, industry, and academia.

The roles of academia and industry would be essentially the same for each research project because of the nature of the role that academia and industry play in the development of new technologies and products.

The FAA would be most directly engaged in the VV&C research project, because certification of civil aviation systems is one of its core functions. However, the subject matter of most of the research projects is related to certification directly or indirectly, so the FAA would be ultimately be interested in the progress and results of those projects.

DOD is primarily concerned with military applications of IA systems, though it must also ensure that military aircraft with IA systems that are based in the United States satisfy requirements for operating in the NAS. Its interests and research capabilities encompass the scope of all eight research projects, especially with regard to the roles of personnel and systems and human cognizance and control.

NASA would support basic and applied research in civil aviation tools, methods, and technologies (including ATM technologies of interest to the FAA) for application in the near, medium, and far term for a wide range of operating concepts. Its interests and research capabilities also encompass the scope of all eight research projects, particularly modeling and simulation, nontraditional methodologies and technologies, and safety and efficiency.

[10] P. Slovic, 1987, Perception of risk, *Science* 236(4799): 280-285.

For each of these organizations, it will be important to involve researchers with relevant expertise who might not normally see themselves as addressing civil aviation issues. For example, most research and development activity related to machine learning, artificial intelligence, and robotics is not taking place in the context of IA systems for civil aviation, but it will be essential to draw on the latest research and development in these areas to most effectively carry out some of the research projects.

Each of the high-priority research projects overlaps to some extent with one or more of the other projects, and each would be best addressed by multiple organizations working in concert. There is already some movement in that direction.

The FAA has created the Unmanned Aircraft Systems Integration Office to foster collaboration with a broad spectrum of stakeholders, including DOD, NASA, industry, academia, and technical standards organizations. The FAA is working with this broad community to support safe and efficient integration of UAS in the NAS. Key activities include the development of regulations, policies, guidance material, and training requirements. In addition, the FAA is coordinating with relevant departments and agencies to address related policy areas such as privacy and national security, and in 2015 it will establish an air transportation center of excellence for UAS research, engineering, and development.[11,12]

The Senior Policy Committee that oversees the work of the NextGen Joint Planning and Development Office is chaired by the Secretary of Transportation and includes representatives from the FAA, NASA, DOD, the Department of Commerce, the Department of Homeland Security, and the White House Office of Science and Technology Policy. The NextGen Program is executing a multiagency research and development plan to improve the NAS. The Senior Policy Committee views the integration of UAS into the NAS as a national priority and, to that end, in 2012 it published the *NextGen Unmanned Aircraft Systems Research, Development and Demonstration Roadmap Version 1.0*. As the title suggests, this document describes an approach for coordinating UAS research, development, and demonstration projects across the agencies involved in NextGen.

There are also positive indications of coordination in the recreational or hobbyist unmanned aircraft community. There are many websites that share open-source software and hardware, with open forums that enable widespread sharing of ideas, techniques, and methods for achieving IA operations of small, amateur-built UAS. However, these activities are largely undertaken without consideration of issues such as certification and thus are generating designs and approaches that might be technologically advanced but could be certified for use in the civil aviation system only with great difficulty, if at all. Since these systems are not verified or validated for qualities such as robustness, completeness, and fault tolerance, the ultimate contribution of this sector to the advancement of IA systems in the NAS remains to be seen.

The collaborative interagency and private efforts described above are necessary and could be strengthened to assure that the full scope of IA research and development efforts (not just those focused on UAS applications) are effectively coordinated and integrated, with minimal duplication of research and without critical gaps. In particular, more effective coordination among relevant organizations in government, academia, and industry would help execute the recommended research projects more efficiently, in part by allowing lessons learned from the development, test, and operation of IA systems to be continuously applied to ongoing activities.

The recommended research agenda would directly address the technology barriers and the regulation and certification barriers. As noted in Table 4.1, although several research projects would address the social and legal issues, the agenda would not address the full range of these issues. In the absence of any other action, resolution of the legal and social barriers will likely take a long time, as court cases are filed to address various issues in various locales on a case-by-case basis, with intermittent legislative action taken in reaction to highly publicized court cases, accidents, and the like. A more timely and effective approach for resolving the legal and social barriers could begin with discussions involving the Department of Justice, FAA, the National Transportation Safety Board, state attorneys general, public interest legal organizations, and aviation community stakeholders. The discussion of some related issues may also be informed by social science research. Given that the FAA is the federal govern-

[11] FAA, 2013, *Integration of Civil Unmanned Aircraft Systems (UAS) in the National Airspace System (NAS) Roadmap,* November.

[12] FAA, Center of Excellence for Unmanned Aircraft Systems, "Welcome," http://faa-uas-coe.net/, accessed April 3, 2014.

ment's lead agency for establishing and implementing aviation regulations, it is in the best position to take the lead in initiating a collaborative and proactive effort to address legal and social barriers.

Executing the research agenda set forth in this report would require significant resources from multiple federal agencies and research organizations. However, substantial advances could be achieved using currently available resources. In large part, the primary responsibilities of specific agencies for given elements of the research agenda will be determined by the resident expertise and specialized research facilities. For example, as noted above, there is one critical, crosscutting challenge that must be overcome to unleash the full potential of advanced IA systems in civil aviation: How can we assure that advanced IA systems—especially those systems that rely on adaptive/nondeterministic software—will enhance rather than diminish the safety and reliability of the NAS? Executing the research projects that will overcome this critical challenge would require a broad mix of expertise, analytic capabilities, and specialized facilities that are resident at research, development, test, and manufacturing organizations within NASA, the FAA, DOD, industry, and academia. It would also require an ongoing commitment of resources and expertise and a willingness to work together toward a shared goal.

CONCLUDING REMARKS

Civil aviation in the United States and elsewhere in the world is on the threshold of profound changes in the way it operates because of the rapid evolution of IA systems. Advanced IA systems will, among other things, be able to operate without direct human supervision or control for extended periods of time and over long distances. As happens with any other rapidly evolving technology, early adapters sometimes get caught up in the excitement of the moment, producing a form of intellectual hyperinflation that greatly exaggerates the promise of things to come and greatly underestimates costs in terms of money, time, and—in many cases—unintended consequences or complications. While there is little doubt that over the long run the potential benefits of IA in civil aviation will indeed be great, there should be equally little doubt that getting there, *while maintaining or improving the safety and efficiency of U.S. civil aviation,* will be no easy matter. Furthermore, given that the potential benefits of advanced IA systems—as well as the unintended consequences—will inevitably benefit some stakeholders much more than others, the enthusiasm of the latter for fielding such systems could be limited. In any case, overcoming the barriers identified in this report by pursuing the research agenda proposed by the committee is a vital next step, although more work beyond the issues identified here will certainly be needed as the nation ventures into this new era of flight.

5

Findings and Recommendation

Finding. *Potential Benefits and Risks.* The intensity and extent of autonomy-related research, development, implementation, and operations in the civil aviation sector suggest that there are several potential benefits to increased autonomy for civil aviation. These benefits include but are not limited to improved safety and reliability, reduced acquisition and operational costs, and expanded operational capabilities. However, the extent to which these benefits are realized will be greatly dependent on the degree to which the barriers that have been identified are overcome, the extent to which military expertise and systems can be leveraged, and the extent to which government and nongovernment efforts are coordinated.

Finding. *Barriers.* There are many substantial barriers to the increased use of autonomy in civil aviation systems and aircraft:

- Technology Barriers[1]
  —Communications and data acquisition,
  —Cyberphysical security,
  —Diversity of aircraft,
  —Human–machine integration,
  —Decision making by adaptive/nondeterministic systems,
  —Sensing, perception, and cognition,
  —System complexity and resilience, and
  —Verification and validation.
- Regulation and Certification Barriers
  —Airspace access for unmanned aircraft,
  —Certification process,
  —Equivalent level of safety, and
  —Trust in adaptive/nondeterministic IA systems.

[1] The committee did not prioritize the barriers; they are listed alphabetically within each group.

- Additional Barriers
  —Legal issues and
  —Social issues.

Finding. ***Development of New Regulations.*** As with the previous introduction of significantly new technologies, such as fly-by-wire and composite materials, the FAA will need to develop technical competency in IA systems and issue new guidance material and regulations to enable safe operation of all classes and types of IA systems.

Recommendation. ***National Research Agenda.*** **Agencies and organizations in government, industry, and academia that are involved in research, development, manufacture, certification, and regulation of IA technologies and systems should execute a national research agenda in autonomy that includes the following high-priority research projects, with the first four being the most urgent and the most difficult:**

- ***Behavior of Adaptive/Nondeterministic Systems.*** **Develop methodologies to characterize and bound the behavior of adaptive/nondeterministic systems over their complete life cycle.**
- ***Operation Without Continuous Human Oversight.*** **Develop the system architectures and technologies that would enable increasingly sophisticated IA systems and unmanned aircraft to operate for extended periods of time without real-time human cognizance and control.**
- ***Modeling and Simulation.*** **Develop the theoretical basis and methodologies for using modeling and simulation to accelerate the development and maturation of advanced IA systems and aircraft.**
- ***Verification, Validation, and Certification.*** **Develop standards and processes for the verification, validation, and certification of IA systems, and determine their implications for design.**
- ***Nontraditional Methodologies and Technologies.*** **Develop methodologies for accepting technologies not traditionally used in civil aviation (e.g., open-source software and consumer electronic products) in IA systems.**
- ***Roles of Personnel and Systems.*** **Determine how the roles of key personnel and systems, as well as related human–machine interfaces, should evolve to enable the operation of advanced IA systems.**
- ***Safety and Efficiency.*** **Determine how IA systems could enhance the safety and efficiency of civil aviation.**
- ***Stakeholder Trust.*** **Develop processes to engender broad stakeholder trust in IA systems for civil aviation.**

Appendixes

A

Statement of Task

The National Research Council will appoint an ad-hoc committee to develop a national research agenda for autonomy in civil aviation, comprised of a prioritized set of integrated and comprehensive technical goals and objectives of importance to the civil aeronautics community and the nation. The elements of the recommended research agenda for autonomy in civil aviation will be evolved from the existing state of the art, scientific and technological requirements to advance the state of the art, potential user needs, and technical research plans, programs, and activities. In addition, the committee will consider the resources and organizational partnerships required to complete various elements of the agenda.

In particular the committee will:

1. Consider the current context of research in autonomy relevant to civil aviation based on factors such as the following:
   a) The current state of the art in autonomy research and applications (for example, national defense, space, automotive, and marine applications) by U.S. industry, NASA, the Department of Defense, the Federal Aviation Administration, other federal agencies, academia, and non-U.S. research agencies and organizations and the contributions that these organizations are making to pursue the development and application of advanced autonomy in relevant civil aviation systems, vehicles (both crewed and unmanned), processes, and mission capabilities.
   b) Current national guidance on research goals and objectives in autonomy in civil aviation.

2. Describe the following:
   a) The scope of the committee's investigation in terms of the forms and applications of autonomy that the committee considered.
   b) Contributions that advances in autonomy could make to civil aeronautics over the next 10 to 20 years through research that provides (i) a steady pace of incremental advances and (ii) credible, game-changing advances in current capabilities.
   c) Technical and policy barriers to implementing advances in autonomy in operational civil aeronautics systems and how those barriers might be overcome.
   d) Key challenges and gaps that a national research agenda in autonomy for civil aviation should address.

3. Outline a potential national research agenda for autonomy in civil aviation, as follows:
   a) The research agenda should consist of a prioritized set of research projects that, if successful,
      i) would enable concepts of operation for the national airspace system where vehicles and systems with various autonomous capabilities are able to operate in harmony with each other and human operators/supervisors,
      ii) could lead to the development, integration, testing, and demonstration of advanced autonomy capabilities for vehicles and systems (both crewed and unmanned), processes, and mission capabilities,
      iii) predict the system-level effects of incorporating the above in the national airspace system, and
      iv) define approaches for verification, validation, and certification of new forms and applications of autonomy.
   b) The agenda should be developed with due consideration of the resources and organizational partnerships required to complete the projects included in the agenda.
   c) For each project, the agenda should, as appropriate, describe the potential contributions and role of U.S. research organizations, including NASA, other federal agencies, industry, and academia.

B

Committee and Staff Biographical Information

JOHN-PAUL B. CLARKE, *Co-chair,* is an associate professor at the Daniel Guggenheim School of Aerospace Engineering with a courtesy appointment in the H. Milton Stewart School of Industrial and Systems Engineering, and is director of the Air Transportation Laboratory at the Georgia Institute of Technology. His research and teaching in the areas of control, optimization, and system analysis, architecture, and design are motivated by his desire to simultaneously maximize the efficiency and minimize the societal costs (especially those imposed on the environment) of the global air transportation system. Dr. Clarke has been recognized globally for his seminal contributions to air traffic management, aircraft operations, and airline operations. His honors include the 1999 AIAA/AAAE/ACC Jay Hollingsworth Speas Airport Award, which is awarded jointly by the American Institute of Aeronautics and Astronautics (AIAA), the American Association of Airport Executives, and the Airports Consultants Council; the 2003 FAA Excellence in Aviation Award, the 2006 National Academy of Engineering Gilbreth Lectureship, and the 2012 AIAA/SAE William Littlewood Lectureship. He is an associate fellow of the AIAA and a member of the Airline Group of the International Federation of Operational Research Societies, the Institute for Operations Research and the Management Sciences, and Sigma Xi. Dr. Clarke has carried out extensive research into air traffic management systems, airline operations, and the air transportation system as a whole. He has an Sc.D. from the Massachusetts Institute of Technology. He is a member of the NRC Committee on Review of the Enterprise Architecture, Software Development Approach, and Safety and Human Factor Design of the Next Generation Air Transportation System and a former member of the Aeronautics and Space Engineering Board, the Committee on Analysis of Air Force Engine Efficiency Improvement Options for Large Non-Fighter Aircraft, and a panel of the Decadal Survey of Civil Aeronautics study.

JOHN K. LAUBER, *Co-chair,* is a private consultant. He previously served as vice president in three different divisions of Airbus and as vice president for corporate safety and compliance at Delta Air Lines. Previously, Dr. Lauber served two terms as a member of the National Transportation Safety Board (NTSB). He is licensed as a commercial pilot with both airplane and helicopter ratings and is type-rated in the B727 and the A320. He has received numerous awards, including NASA's Outstanding Leadership Award and the Joseph T. Nall Award from the International Aviation and Transportation Safety Bar Association. He has served as president of the International Federation of Airworthiness and the Association for Aviation Psychology. Dr. Lauber is nominated to this committee primarily because of his extensive background in operational safety from the perspective of a government safety organization (the NTSB), an airline (Delta), and an aircraft manufacturer (Airbus). Dr. Lauber holds a Ph.D. in neuropsychology from Ohio State University. He is a former member of the NRC's Aeronautics and Space Engineering Board, the Air Force Studies Board, and seven NRC committees and panels, including (most recently) the Committee on the Effects of Commuting on Pilot Fatigue and the Committee on Technology Pathways: Workshops to Review the Integrated Plan for a Next Generation Air Transportation System.

BRENT APPLEBY is the deputy to the vice president of engineering for science and technology at Draper Laboratory. At Draper, Dr. Appleby is responsible for shaping the Lab's technology development strategy, supporting external science and technology marketing efforts, and external outreach to universities for advanced technology collaborations. Former positions at Draper include Mission Systems Division lead, Tactical ISR Division lead, and Advanced Control Systems group leader. Dr. Appleby also recently completed an assignment at the Defense Advanced Research Projects Agency (DARPA), where he was the deputy director of the Strategic Technology Office. He participated as a member of a recent Defense Science Board task force that released a report in 2012, *The Role of Autonomy in Department of Defense Systems*. Dr. Appleby has also been a lecturer in the Aero/Astro Engineering Department at MIT and has supervised many MIT graduate students in their research efforts. His main technical background is in guidance, navigation, and control system technology, and he has worked on numerous air, space, and undersea crewed and unmanned systems in his career. Dr. Appleby has experience in strategic planning for development of advanced technologies, as well as in a broad spectrum of unmanned vehicles. Dr. Appleby has a Ph.D. in aeronautics and astronautics from the Massachusetts Institute of Technology.

ELLA M. ATKINS is an associate professor in the Department of Aerospace Engineering at the University of Michigan. She is also director of the Autonomous Aerospace Systems Laboratory. She previously served on the aerospace engineering faculty at the University of Maryland, College Park. Dr. Atkins' research focuses on the integration of strategic and tactical planning and optimization algorithms to enable robust operation in the presence of system failures and environmental uncertainties. She has collaboratively pursued challenging autonomous flight applications for crewed aircraft and UAS, including the Flying Fish autonomous unmanned seaplane. Dr. Atkins also studies the optimization of and safety analysis in congested airspace, with early efforts in simultaneous noninterfering terminal area airspace planning for runway-independent aircraft and ongoing research in safety assessments for small unmanned aircraft during low-altitude flight operations. Comprehensive aerodynamic sensing has been explored for small flapping and fixed-wing UAS to investigate their use to improve controllability in commanded or unintentional outside-the-envelope flight conditions. She is author of more than 100 journal and conference publications and serves as an associate editor for the *AIAA Journal of Aerospace Information Systems* (formerly the *AIAA Journal of Computing, Information, and Communication*). Dr. Atkins is past chair of the AIAA Intelligent Systems Technical Committee, an associate fellow of AIAA, and a senior member of the Institute of Electrical and Electronics Engineers (IEEE). She is also a owner/operator of small public airport (Shamrock Field, Brooklyn, Michigan) and is a private pilot (airplane, single engine, land) and an Academy of Model Aeronautics pilot (radio control). She has expertise in developing robust, fault-tolerant air transportation systems to enhance the safety of crewed and unmanned aircraft. Dr. Atkins has a Ph.D. in computer science and engineering from the University of Michigan. She is a member of the NRC's Aeronautics and Space Engineering Board and of the Aeronautics Research and Technology Roundtable and is a former member of the Committee for the Review of NASA's Aviation Safety Related Programs and of Panel E: Intelligent and Autonomous Systems, Operations and Decision Making, Human Integrated Systems, Networking, and Communications as part of the Decadal Survey of Civil Aeronautics.

ANTHONY J. BRODERICK is an independent consultant specializing in aviation safety. His clients include domestic and international airlines, aerospace firms, a major aircraft manufacturer, and governments. He sits on the technical advisory board of the Center for Advanced Aviation System Development at the MITRE Corporation, and from 2002 to 2014 he sat on the Advisory Council of the Institute of Nuclear Power Operations. Upon his retirement in June 1996 from his post as associate administrator for regulation and certification in the Federal Aviation Administration, he had been for more than 11 years the senior career aviation safety official in the federal government. As head of the FAA's Regulation and Certification organization, he was principally responsible for the development and enforcement of policy and regulations governing the certification, production approval, and continued airworthiness of aircraft; the certification of pilots, mechanics, and others in safety-related positions; the certification of all operational and maintenance enterprises engaged in U.S. civil aviation, both domestic and overseas; development of regulations; civil flight operations; and the certification and safety oversight of some 7,300 U.S. commercial airlines and air operators. Mr. Broderick led the agency's development of the International Aviation Safety Assessment program, which in 1996 became the model for the International Civil Aviation Organization safety assessment program. He was also instrumental in leading international efforts to establish certification and operational standards for safely allowing extended-range twin-engine airliner operations; early operational implementation of the Global Positioning System; and harmonization of certification, operations, and maintenance standards, among many other safety initiatives. He has a B.S. in physics from St. Bonaventure University. He has been a member of the NRC's Committee on NASA's National Aviation Operational Monitoring Service (NAOMS) Project: An Independent Assessment; the Aeronautics and Space Engineering Board; and the Panel on Transportation as part of the study on Science and Technology for Countering Terrorism. He has also chaired the NRC's Committee to Conduct an Independent Assessment of the Nation's Wake Turbulence Research and Development Program.

GARY L. COWGER (NAE) is chairman and CEO of GLC Ventures, LLC, a consulting firm he founded after retiring from General Motors Corporation (GMC) as the group vice president for manufacturing and labor. He previously held a variety of other positions at GM, including president and managing director of GM de Mexico; manufacturing vice president for GM Europe; chairman, Adam Opel AG; and president, GM North America. Mr. Cowger has extensive experience in benchmarking, target setting, and creating and applying organizational and production-based performance measures. Mr. Cowger holds an M.S. in management from the Massachusetts Institute of Technology and a B.S. in industrial engineering. His recent service on NRC committees includes the Committee on the Potential for Light-Duty Vehicle Technologies 2010-2050 and the Industrial, Manufacturing and Operational Systems Engineering Search Committee.

CHRISTOPHER E. FLOOD is captain and check airman for the Boeing 737-700/800/900 at Delta Air Lines, where he is also a member of the Aviation Safety Action Program (ASAP) Event Review Committee and has served as a member of several line operations safety audit working groups. Previous positions at Delta Air Lines include system manager for labor relations (responsibilities included labor relations and working agreements with 10,000 Delta pilots); Salt Lake City chief pilot (responsibilities included leadership of 700 pilots, safety of flight operations at 21 airports, and investigation of operational incidents); and line pilot. Captain Flood has also participated in Delta Air Lines' fleet evaluation of Airbus vs. Boeing aircraft; development and implementation of standard terminal arrival routes, departure procedures, and airspace enhancements; and management of flight operations at Delta's European hub. While associated with Delta Air Lines, Captain Flood has concurrently worked as an attorney with The Aviation Law Firm, where he assisted in the representation of clients in enforcement actions before the National Transportation Safety Board and the Department of Transportation; provided consulting and legal services to air carriers operating under Part 135 of the Federal Aviation Regulations; and represented survivors of three medevac helicopter accidents in personal injury lawsuits. Captain Flood began his aviation career in the U.S. Air Force, where he served as a pilot, weapons officer, and instructor pilot. He is a member of the District of Columbia Bar and the Flight Safety Foundation, and he is also an FAA Safety Team Representative. Captain Flood has a J.D. from the University of Cincinnati.

MICHAEL S. FRANCIS is the chief, Advanced Programs, and a senior fellow at the United Technologies Research Center (UTRC), leading its initiative in autonomous and intelligent systems. He also serves as the program executive for autonomous systems at Sikorsky Innovations, guiding Sikorsky's entry into the optionally piloted and unmanned aeronautical systems market. Dr. Francis's involvement with research and development of unmanned air systems dates to the early 1990s, when he initiated the Unmanned Tactical Aircraft Program, which later became the Unmanned Combat Air Vehicle Program at DARPA. During that period, he also initiated the agency's first Micro Air Vehicle Program and directed the award-winning U.S.-German X-31 International Experimental Fighter

Aircraft Program through flight test. He was also the DARPA program director for the $4 billion DARPA-Air Force-Navy joint Unmanned Combat Air Systems Office. Dr. Francis's 27-year military career spanned the spectrum of aviation and space research and development, beginning with his assignment to the U.S. Air Force Academy as a research scientist and professor. He later served as a program manager for advanced aerodynamics research and development at the Air Force Office of Scientific Research. There, he established the Air Force's first turbulent flow control program. He went on to serve at the Air Force Space and Missile Systems Center, where he managed a variety of advanced space system development efforts, including classified programs, for the Air Force and the Strategic Defense Initiative Organization. In his last military assignment as the integrator and architect for the Pentagon's Defense Airborne Reconnaissance Office, then Col. Francis led the development of the first Department of Defense integrated air–space surveillance architecture in collaboration with the National Reconnaissance Office. Following his military retirement in 1997, Dr. Francis became the first president of Athena Technologies, then a small start-up company specializing in advanced digital control systems for robotic applications. Dr. Francis later joined Lockheed Martin as an executive at its corporate headquarters, leading a team that developed cross-corporation and strategic initiatives. Immediately prior to joining UTRC, he served as the chief operating officer of the Photonics Division of General Atomics. Dr. Francis is a fellow of the AIAA. He has a Ph.D. in aerospace engineering sciences as well as an honorary doctoral degree from the University of Colorado.

ERIC FREW is an associate professor and director of the Research and Engineering Center for Unmanned Vehicles at the University of Colorado, Boulder. He is also the University of Colorado site director for the Center for Unmanned Aircraft Systems (CUAS), a National Science Foundation Industry/University Cooperative Research Center. Dr. Frew has led the development and field deployment of a variety of unmanned aircraft systems for communication and sensing applications. He was one of the lead members of the team that performed the first-ever sampling of the rear flank gust front of a tornadic supercell thunderstorm using a small unmanned aircraft system. Dr. Frew's research efforts focus on autonomous flight of heterogeneous unmanned aircraft systems; optimal distributed sensing by mobile robots; controlled mobility in ad hoc sensor networks; miniature self-deploying systems; and guidance and control of unmanned aircraft in complex atmospheric phenomena. He received the National Science Foundation Faculty Early Career Development Award in 2009 and was selected for the 2010 DARPA Computer Science Study Group. He has a Ph.D. in aeronautics and astronautics from Stanford University.

ANDREW LACHER is the UAS integration research lead and a senior principal with the MITRE Corporation. He is responsible for strategic coordination of MITRE's research associated with the integration of UAS into civil airspace and with the challenges of integrating complex automation technologies in the Next Generation Air Transportation System (NextGen). Mr. Lacher is part of a cross-corporate team exploring the implications of autonomy for MITRE's broad sponsor base. Mr. Lacher has over 25 years of system engineering, management, and strategic research planning experience in a variety of aviation domains. He was heavily involved with the collaborative decision-making initiative that established a real-time link between airline operational control and the FAA's Traffic Flow Management office. Mr. Lacher was part of the implementation of the FAA's Joint Planning and Development Office and the initial definition of NextGen, a major, decades-long improvement to the U.S. air transportation system that will include a much higher degree of automation than the current air transportation system. He has also worked as a strategic airline consultant responsible for the planning and integration of the information technical infrastructure to support system operations, flight dispatch and release, weight and balance, maintenance, crew scheduling and planning, and performance monitoring. He was manager of customer integration for ORBCOMM, Inc. Mr. Lacher has an M.S. in operations research from the George Washington University. He is a member of the NRC's Aeronautics Research and Technology Roundtable and a former member of Panel E: Intelligent and Autonomous Systems, Operations and Decision Making, Human Integrated Systems, Networking, and Communications for the Decadal Survey of Civil Aeronautics.

JOHN D. LEE is the Emerson Professor in the Department of Industrial and Systems Engineering at the University of Wisconsin–Madison. Previously he was with the University of Iowa, where he was the director of human factors research at the National Advanced Driving Simulator. Before moving to the University of Iowa, he was a research scientist at the Battelle Human Factors Transportation Center for six years. His research focuses on the safety and acceptance of complex human–machine systems by considering how technology mediates attention. Specific research interests include trust in technology, advanced driver assistance systems, and driver distraction. He is a coauthor of the textbook *An Introduction to Human Factors Engineering* and the author or coauthor of 170 articles. Dr. Lee recently helped edit the book *Driver Distraction: Theory, Effects, and Mitigation*. Dr. Lee serves on the editorial boards of *Cognitive Engineering and Decision Making; Cognition, Technology and Work;* and *International Journal of Human Factors Modeling and Simulation* and is the associate editor for the journals *Human Factors* and *IEEE-Systems, Man, and Cybernetics*. He received the Ely Award for best paper in the journal *Human Factors* (2002) and the best paper award for the journal *Ergonomics* (2005). Dr. Lee has a Ph.D. in mechanical engineering from the University of Illinois at Urbana-Champaign. He has been a member of seven NRC study groups, most recently the Committee on Electronic Vehicle Controls and Unintended Acceleration, the Committee on Naval Engineering in the 21st Century, and the Committee on Human-Systems Integration.

KENNETH M. ROSEN, NAE, is president of General Aero-Science Consultants LLC. He is also a founding principal partner of Aero-Science Technology Associates, LLC, which is an engineering and business development consulting firm established to service both government and industry customers. Before founding General Aero-Science Consultants in 2000, Dr. Rosen spent 38 years with the United Technologies Corporation, primarily with Sikorsky Aircraft. Dr. Rosen held many major engineering and management positions at Sikorsky, including vice president of research and engineering and vice president of advanced programs and processes. He directed all the company's advanced-technology military and commercial projects, including the Comanche helicopter, the S-92 helicopter (which won the 2003 Collier trophy), the Cypher unmanned air vehicle, the UH-60 Black Hawk helicopter, the SH-60 Seahawk helicopter, the S-76 helicopter, and the X-wing helicopter. During that time Dr. Rosen managed all of Sikorsky's research, systems engineering, product development, design, production engineering, ground and flight test, and avionics and systems integration efforts. Dr. Rosen was a member of the Sikorsky Executive Board and was also responsible for all of the company's advanced products and low observable activities. He has served the DARPA Tactical Technology Office as a senior advisor supporting advanced aerospace research programs such as the unmanned Combat Armed Rotorcraft, Heliplane, and Heavy Lift Helicopter. Recently he helped prepare the Future of Vertical Lift Aviation study for the U.S. Army and DARPA. He is an elected member of the National Academy of Engineering and the Connecticut Academy of Science and Engineering, as well as a fellow of the American Society of Mechanical Engineers, the Royal Aeronautical Society, the Society of Automotive Engineers, the American Institute of Aeronautics and Astronautics, and the American Helicopter Society. Dr. Rosen is a recipient of the NASA Civilian Public Service Medal, the Dr. Alexander Klemin Award for lifetime achievement from the American Helicopter Society, and Vice President Al Gore's "Hammer" award from the Department of Defense for innovative cost management. He has been chairman of the board of the Rotorcraft Industry Technology Association, chairman of the United Technologies Corporation Engineering Coordination Steering Committee, vice chairman of the Software Productivity Consortium, and chairman of the Aerospace Industries Association Rotorcraft Advisory Group. He has also been a member of NASA's Aeronautics and Space Transportation Technology Advisory Committee and the SAE Aerospace Council. He holds five U.S. patents and has authored numerous papers in the fields of helicopter design, tilt rotor optimization, product development, propulsion, aerothermodynamics, icing, and systems engineering. Dr. Rosen has a Ph.D. in mechanical engineering from Rensselaer Polytechnic Institute and is a graduate of the Advanced Management Program at the Harvard University Business School. Dr. Rosen has served as a member of the NRC's Panel on Mechanical Science and Engineering at the Army Research Laboratory and the Panel on Air and Ground Vehicle Technology.

LAEL RUDD is the autonomy development lead at Northrop Grumman Aerospace Systems, a major developer and manufacturer of unmanned air vehicles, including the Global Hawk. He has led the technical efforts for various aircraft and military space programs and proposals, providing control laws to assist with development of a human-in-the-loop test-bed. Prior to working at Northrop Grumman, Dr. Rudd worked for the Aerospace Corporation, first in the Guidance and Controls Department and later in the Flight Software Validation Department. Dr. Rudd is an associate fellow of AIAA, serving on the Guidance, Navigation, and Control Technical Committee, and a senior member of IEEE. He has been a part-time lecturer at University of California, Los Angeles, and the University of Southern California; he holds two patents and eight trade secrets within Northrop Grumman and has published technical papers on hypersonic vehicle stability and control, trajectory optimization, adaptive control, and control of both micro-air vehicles and spacecraft. He has a Ph.D. in aerospace engineering from the University of Maryland, College Park.

PATRICIA VERVERS is an engineer fellow at Honeywell Aerospace. She works in the Human Centered Systems Group in Honeywell's Advanced Technology Group in Columbia, Maryland. Since joining Honeywell, Dr. Ververs has led programs in the areas of high-speed research, flight-critical systems research, the System Wide Accident Prevention Project of NASA's Aviation Safety Program, advanced primary flight displays, improving information intake under stress and/or augmented cognition, neurotechnology for information analysts, synthetic and enhanced vision systems, combined vision systems, and advanced features for commercial helicopters. Dr. Ververs's areas of interest include avionics display design, alerting, and notification; cognitive state assessment; and human factors evaluation. Currently, she serves as the technical lead for the Honeywell Combined Vision Systems Program, which is designing and evaluating a next-generation avionics display that integrates real-time infrared imagery with Honeywell's innovative synthetic vision technology for advanced cockpits. The Combined Vision Systems research team received the 2011 Honeywell Corporate Innovation award. She also leads a program to bring similar display technology to commercial helicopters to provide pilots with situation awareness of the surrounding terrain and obstacles. Dr. Ververs currently serves as the U.S.-based staff scientist on the European Union's Single European Sky Air Traffic Management Research program for head-up and head-down synthetic and combined vision solutions. She is also the human factors lead on the Boeing team for the FAA Systems Engineering 2020 program. Dr. Ververs regularly serves on RTCA Special Committees establishing minimum aviation system performance standards for new avionics equipment. She holds seven patents on alerting and notification, display design, and workload monitoring and has filed more than a dozen other patents in the areas of display design and the human–machine interface. She is a five-time recipient of Honeywell Aerospace's Technical Award. She received *Avionics Magazine*'s Women in Technology Emerging Leader Award in 2012 and the 1999 Stanley N. Roscoe award from the Human Factors and Ergonomics Society for the best doctoral dissertation in the area of aerospace human factors She has a Ph.D. in engineering psychology from the University of Illinois at Urbana-Champaign.

LARRELL B. WALTERS is the head of the Sensor Systems Division of the University of Dayton Research Institute, a division that has grown from 3 to more than 75 people in just over 6 years. The scope of the division's work encompasses wide-area situational awareness, image processing and exploitation algorithms, visualization, human factors, and unmanned autonomous systems. Mr. Walters is also the director of the Institute for the Development and Commercialization of Advanced Sensor Technology, which is a collaborative research entity involving eight universities and over 20 companies. Previously Mr. Walters worked in commercial aerospace for 20 years with Goodyear and Goodrich to provide advanced technical solutions and aircraft systems for the airline industry. He led technical product support and sales teams before becoming vice president of operations for a large business unit that overhauled a wide array of items removed from commercial aircraft. Mr. Walters subsequently moved into the field of microelectronics, serving as president of two companies that designed and built complex circuitry and developed capital assets that supported the manufacture of silicon-based electrical devices. He has received several awards, most recently the Award for Excellence in Technical Leadership from the Affiliate Societies Council Dayton and the Honorary Alumnus Award from the Sinclair Community College for helping them establish an extensive program for unmanned autonomous systems education and certification. Mr. Walters has a B.S. in computer science from Bowling Green State University and an M.B.A. from Kent State University.

DAVID WOODS is a professor in the Department of Integrated Systems Engineering and the Institute for Ergonomics at the Ohio State University. He also leads the university-wide Initiative on Complexity in Natural, Social, and Engineered Systems. A pioneer in cognitive systems engineering for human–computer decision making in emergencies, Dr. Woods studies the brittleness and resilience of human-automation systems—in crisis response, in nuclear power emergencies, in pilot-automation teams, in anomaly response in space shuttle mission operations, in critical care medicine, in professional information analysis, and with robots (unmanned air vehicles and unmanned ground vehicles) in military and disaster response missions. He has investigated accidents in nuclear power, aviation, space, and anesthesiology and was an advisor to the Columbia Accident Investigation Board. A new direction in his research on safety is how to engineer resilience into systems that manage high-risk processes. As a pioneer in developing this new direction, he is coeditor of two books (*Resilience Engineering*, 2006, and *Resilience Engineering in Practice*, 2011) and 20 publications on this topic. His latest book in preparation is *Outmaneuvering Complexity*. Other books include *Behind Human Error* (1994; 2nd edition 2010), *Joint Cognitive Systems: Foundations of Cognitive Systems Engineering* (2005), and *Joint Cognitive Systems: Patterns in Cognitive Systems Engineering* (2006). Dr. Woods is currently president of the Resilience Engineering Association, past president and a fellow of the Human Factors and Ergonomics Society, and a fellow of the Association for Psychological Science and the American Psychological Association. He is corecipient of the Ely Award for best paper in the journal *Human Factors* (1994). He has also received the Laureate Award from *Aviation Week and Space Technology* (1995) for research on the human factors of highly automated cockpits, the Jack Kraft Innovators Award from the Human Factors and Ergonomics Society (2002), and the Jimmy Doolittle Fellow award for Research on Human Performance from the Air Force Association (2012). Dr. Woods has served as a member of the FAA's Human Factors Study Team on Advanced Flight Decks (1996), Aerospace Research Needs (2003), and Dependable Software (2006). He participated as a member of a recent Defense Science Board task force that released a report in 2012 titled *The Role of Autonomy in DoD Systems*, and he has testified to the U.S. Congress on safety at NASA and on election reform. Dr. Woods has a Ph.D. in cognitive psychology from Purdue University. He has served as a member on the NRC's Committee on Certifiably Dependable Software Systems, the Committee on Engineering and the Health Care System, the Committee on Aeronautics Research and Technology for Vision 2050, and the Panel on Human Factors Research Needs in Nuclear Regulatory Research.

EDWARD L. WRIGHT, NAS, holds the David Saxon Presidential Chair in Physics and a professor of astronomy at the University of California, Los Angeles. Dr. Wright's research interests are in theoretical and experimental infrared astronomy and cosmology, especially cosmic microwave background radiation studies. He played a major role on the NASA Cosmic Background Explorer (COBE) mission, and in 1992 he received the NASA Exceptional Scientific Achievement Medal for this work. He was a coinvestigator on NASA's Wilkinson Microwave Anisotropy Probe, a mission that followed up the COBE discovery of fluctuations in the early universe. Dr. Wright participated in the Joint Efficient Dark-energy Investigation and he was an interdisciplinary scientist on NASA's Spitzer Space Telescope Science Working Group. Dr. Wright is the principal investigator for the Wide-field Infrared Survey Explorer MidEx mission launched in 2009. He has a Ph.D. in astronomy from Harvard University. He has been a member of the NRC's Committee on NASA's Beyond Einstein Program: An Architecture for Implementation; the Panel on Astronomy and Astrophysics; the Committee on Physics of the Universe; and the Panel on Ultraviolet, Optical, and Infrared Astronomy from Space.

STAFF

ALAN C. ANGLEMAN, *Study Director,* has been a senior program officer for the Aeronautics and Space Engineering Board (ASEB) since 1993, directing studies on the modernization of the U.S. air transportation system, system engineering and design systems, aviation weather systems, aircraft certification standards and procedures, commercial supersonic aircraft, the safety of space launch systems, radioisotope power systems, cost growth of NASA Earth and space science missions, and other aspects of aeronautics and space research and technology. Previously, Mr. Angleman worked for consulting firms in the Washington area, providing engineering support services to the DOD and NASA Headquarters. His professional career began with the U.S. Navy, where he served for 9 years as a nuclear-trained submarine officer. He has a B.S. in engineering physics from the U.S. Naval Academy and an M.S. in applied physics from the Johns Hopkins University.

LEWIS B. GROSWALD is an associate program officer for the Space Studies Board (SSB). Mr. Groswald is a graduate of George Washington University, where he received a master's degree in international science and technology policy and a bachelor's degree in international affairs, with a double concentration in conflict and security and Europe and Eurasia. Following his work with the National Space Society during his senior year as an undergraduate, Mr. Groswald decided to pursue a career in space policy, with a focus on educating the public on space issues and formulating policy. He has worked on NRC reports covering a wide range of topics, including near-Earth objects, orbital debris, life and physical sciences in space, and planetary science.

LINDA WALKER has been with the National Academies since 2007 and is now program coordinator for the Board on Physics and Astronomy. Before her assignment with the SSB, she was on assignment with the National Academies Press. Prior to her working at the National Academies, she was with the Association for Healthcare Philanthropy in Falls Church, Virginia. Ms. Walker has 34 years of administrative experience.

ANESIA WILKS joined the SSB as a program assistant in August 2013. Ms. Wilks brings experience working in the National Academies conference management office as well as other administrative positions in the D.C. metropolitan area. She has a B.A. in psychology, magna cum laude, from Trinity University in Washington, D.C.

MICHAEL MOLONEY is the director for Space and Aeronautics at the Space Studies Board and the Aeronautics and Space Engineering Board of the National Research Council of the U.S. National Academies. Since joining the ASEB/SSB, Dr. Moloney has overseen the production of more than 40 reports, including four decadal surveys—in astronomy and astrophysics, planetary science, life and microgravity science, and solar and space physics—a review of the goals and direction of the U.S. human exploration program, a prioritization of NASA space technology roadmaps, as well as reports on issues such as NASA's Strategic Direction, orbital debris, the future of NASA's astronaut corps, and NASA's flight research program. Before joining the SSB and ASEB in 2010, Dr. Moloney was associate director of the Board on Physics and Astronomy (BPA) and study director for the decadal survey for astronomy and astrophysics (Astro2010). Since joining the NRC in 2001, Dr. Moloney has served as a study director at the National Materials Advisory Board, the BPA, the Board on Manufacturing and Engineering Design, and the Center for Economic, Governance, and International Studies. Dr. Moloney has served as study director or senior staff for a series of reports on subject matters as varied as quantum physics, nanotechnology, cosmology, the operation of the nation's helium reserve, new anti-counterfeiting technologies for currency, corrosion science, and nuclear fusion. In addition to his professional experience at the National Academies, Dr. Moloney has more than 7 years' experience as a foreign service officer for the Irish government—including serving at the Irish Embassy in Washington and the Irish Mission to the United Nations in New York. A physicist, Dr. Moloney did his Ph.D. work at Trinity College Dublin in Ireland. He received his undergraduate degree in experimental physics at University College Dublin, where he was awarded the Nevin Medal for Physics.

C

Acronyms

ADS-B	Automatic Dependent Surveillance–Broadcast
ATM	air traffic management
DOD	Department of Defense
EPA	Environmental Protection Agency
FAA	Federal Aviation Administration
FAR	Federal Aviation Regulation
GA	general aviation
GPS	Global Positioning System
IA	increasingly autonomous
ICAO	International Civil Aviation Organization
IFR	instrument flight rules
NAS	National Airspace System
NextGen	Next Generation Air Transportation System
NHTSA	National Highway Traffic Safety Administration
NITRD	Networking and Information Technology Research and Development
OODA	Observe, Orient, Decide, Act
PARS	people, (software) agents, robots, and sensors
TCAS	Traffic Alert and Collision Avoidance System
UAS	unmanned aircraft system(s)

V&V	verification and validation
VFR	visual flight rules
VHF	very high frequency
VV&C	verification, validation, and certification